Classical and Quantum Parametric Phenomena

Classical and Quantum Parametric Phenomena

ALEXANDER EICHLER AND ODED ZILBERBERG

ETH Zurich and University of Konstanz

OXFORD
UNIVERSITY PRESS

Great Clarendon Street, Oxford, OX2 6DP,
United Kingdom

Oxford University Press is a department of the University of Oxford.
It furthers the University's objective of excellence in research, scholarship,
and education by publishing worldwide. Oxford is a registered trade mark of
Oxford University Press in the UK and in certain other countries

Published in the United States of America by Oxford University Press
198 Madison Avenue, New York, NY 10016, United States of America

British Library Cataloguing in Publication Data
Data available

Library of Congress Control Number: 2023935805

ISBN 9780192862709

DOI: 10.1093/oso/9780192862709.001.0001

Printed and bound by
CPI Group (UK) Ltd, Croydon, CR0 4YY

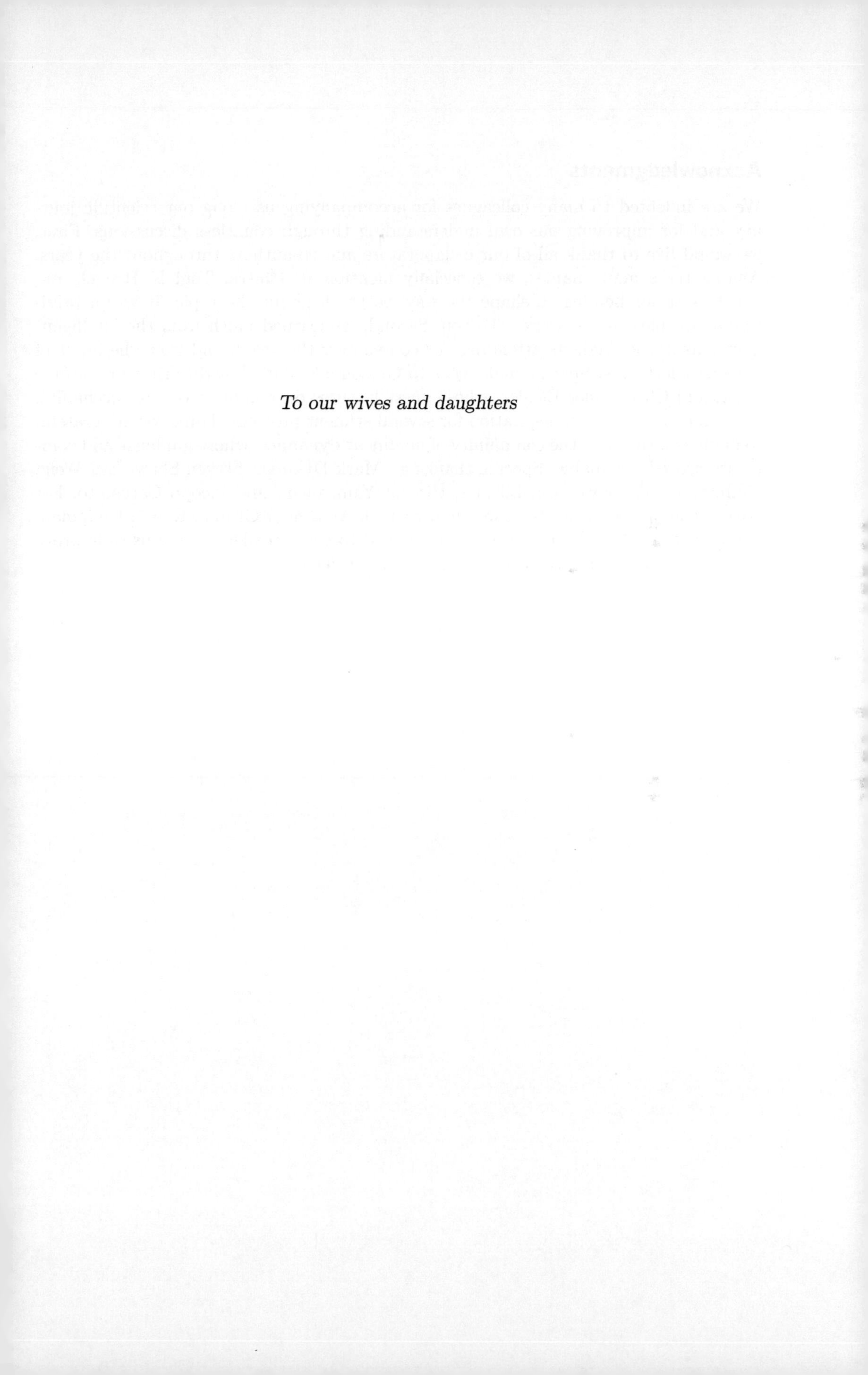

To our wives and daughters

Acknowledgments

We are indebted to many colleagues for accompanying us along our scientific journey and for improving our own understanding through countless discussions. First, we would like to thank all of our collaborators and co-authors throughout the years. Among these many names, we especially mention R. Chitra, Toni L. Heugel, and Jan Košata for helping to shape the way we think about the topic. It was a privilege and a pleasure to work with you. Second, we learned much from the intelligent questions of the students attending our course over the years, and from the input of our guest lecturers. Special credit goes to Giacomo Scalari, Martin Frimmer, Jérôme Faist, and Christopher Eichler, whose guest lectures on nonlinear optics, mechanics, and electronics were an inspiration for several student projects. Third, we are grateful to many colleagues in the community of nonlinear dynamics, whose guidance and comments proved invaluable. Specific thanks go Mark Dykman, Steven Shaw, Eva Weig, Guillermo Villanueva, Ron Lifshitz, Hiroshi Yamaguchi, and Iacopo Carusotto. For critical comments on this text, we further thank Alexander Grimm, Robert Chapman, and Javier del Pino. Finally, we will be obliged to every reader helping us to improve this book by providing constructive criticism and feedback.

Preface

This book provides an overview of the phenomena arising when parametric pumping is applied to oscillators. These phenomena include parametric amplification, noise squeezing, spontaneous symmetry breaking, activated switching, cat states, and synthetic Ising spin lattices. To understand these effects, we introduce topics such as nonlinear and stochastic dynamics, mode coupling, and quantum mechanics. Throughout the book, we keep these introductions as succinct as possible and focus our attention on understanding parametric oscillators. As a result, we familiarize ourselves with many aspects of parametric systems and understand the common theoretical origin of nanomechanical sensors, optical amplifiers, and superconducting qubits.

Parametric phenomena have enabled important scientific breakthroughs over recent decades and are still the focus of intense research efforts. Our intention is to provide a resource for experimental and theoretical physicists entering the field or wishing to gain a deeper understanding of the underlying connections. As such, we combine formal and intuitive explanations, accompanied by exercises based on numerical python codes. This combination allows the reader to experience parametric phenomena from various directions and to apply their understanding directly to their own research projects. For lecturers, the book supplies all the material necessary for an advanced class on the topic.

Preface

Contents

Introduction 1
- 0.1 Historical Review 2
- 0.2 Present and Future 3

1 The Harmonic Resonator 5
- 1.1 Newton's Equation of Motion 5
- 1.2 Response of the Driven Resonator 7
- 1.3 Matrix Formulation 8
- 1.4 Parametric Modulation 11
- 1.5 Floquet Theory 13
- Chapter summary 18
- Exercises 19

2 The Duffing Resonator 20
- 2.1 The Quartic Potential 20
- 2.2 The Cubic Potential 27
- Chapter summary 29
- Exercises 30

3 Degenerate Parametric Pumping 32
- 3.1 The Nonlinear Parametric Resonator 32
- 3.2 Parametric Pumping via Three-Wave Mixing 42
- Chapter summary 44
- Exercises 45

4 Dissipation and Force Fluctuations 46
- 4.1 The Role of Force Noise 46
- 4.2 The Fluctuation–Dissipation Theorem 52
- 4.3 The Probability Distribution Approach 57
- Chapter summary 61
- Exercises 62

5 Parametric Resonators with Force Noise 63
- 5.1 Multistability and Quasi-Stable Solutions 63
- 5.2 Parametric Amplification below Threshold 64
- 5.3 Parametric Pumping Above Threshold 68
- 5.4 Hierarchy of Relevant Timescales 74
- Chapter summary 76
- Exercises 77

6 Coupled Harmonic Resonators 79
6.1 Static Coupling 79
6.2 Nondegenerate Three-Wave Mixing 86
6.3 Alternative Types of Coupling 91
Chapter summary 96
Exercises 97

7 Coupled Parametric Oscillators 98
7.1 Equations for N Coupled Parametric Oscillators 98
7.2 Examples for $N = 2$ 100
7.3 Networks with $N > 2$ 108
Chapter summary 111
Exercises 112

8 The Quantum Harmonic Oscillator 113
8.1 From Classical to Quantum Fluctuations 113
8.2 From First to Second Quantization 116
8.3 Quantum State Representations 123
Chapter summary 130
Exercises 131

9 From Closed to Open Quantum Systems 132
9.1 Coupling to a Thermal Environment 132
9.2 The Driven Quantum Resonator 136
Chapter summary 141
Exercises 142

10 The Quantum Parametric Oscillator 143
10.1 General Hamiltonian 143
10.2 Quantum Parametric Phenomena 145
10.3 Coupled Quantum Parametric Oscillators 153
Chapter summary 156
Exercises 157

11 Experimental Systems 159
11.1 Mechanical Resonator Example 160
11.2 Electrical Resonator Example 161
11.3 Optical Resonator Example 163
11.4 Rescaling of the Numerical Values 166
Chapter summary 168
Exercises 169

References 170

Subject Index 176

List of Important Symbols

(in approximate order of appearance)

H	Hamiltonian
E_{pot}	potential energy
E_{kin}	kinetic energy
E_{tot}	$= E_{\text{pot}} + E_{\text{kin}}$; total energy
x	displacement
p	momentum
t	time
m	mass
ω_0	angular resonance frequency
ν_0	$= \omega_0/(2\pi)$; temporal resonance frequency
k	$= m\omega_0^2$; spring constant
T_0	$= 2\pi/\omega_0 = 1/\nu_0$; unforced oscillator period
Q	quality factor
Γ	$= \omega_0/Q$; damping rate
τ_0	$= 2/\Gamma$; amplitude decay time constant
ω_Γ	$= \left(\omega_0^2 - \Gamma^2/4\right)^{1/2} \approx \omega_0$; dissipation-shifted angular resonance frequency
μ	characteristic exponent
F_0	amplitude of external force
F	in Chapters 1 to 7: all force terms acting on the bare resonator
	in Chapters 9 and 10: $= \frac{F_0}{2}\sqrt{\hbar/2m\omega_0}$; rotating-frame quantum force term
ω	angular frequency of external force
θ	phase offset of external force
χ	susceptibility function of driven resonator
X	oscillation amplitude
$\mathbf{x}$	vector of a system's degrees of freedom
G	in Chapters 1 to 7: matrix containing the coefficients of the differential equation
	in Chapter 10: parametric drive in the rotating-frame quantum Hamiltonian
W	in Chapters 1 to 7: Wronskian matrix
	in Chapters 8 to 10: Wigner quasiprobability density
Φ	state transfer matrix
T_p	period of parametric pump
ω_p	$= 2\pi/T_p$; angular frequency of parametric pump
λ	parametric modulation depth
λ_{th}	$= 2/Q$; parametric pumping threshold at $\omega_p = 2\omega_0$
β_3	coefficient of cubic (Duffing) nonlinearity
β_2	coefficient of quadratic nonlinearity
β	$= \beta_3 - \frac{10}{9}\frac{\beta_2^2}{\omega^2}$; coefficient of effective Duffing nonlinearity
u	in-phase oscillation quadrature
v	out-of-phase oscillation quadrature

ψ	phase offset of parametric pump
η	coefficient of nonlinear damping
k_B	$\approx 1.38 \times 10^{-23}\,\mathrm{J\,T^{-1}}$; Boltzmann constant
T	temperature
E_{eq}	equilibrium energy
ξ	force noise term
ς_D	standard deviation of force noise
σ_x	standard deviation of x (for any variable x)
S_F	power spectral density of force noise
Ξ_u	in-phase quadrature of force noise
Ξ_v	out-of-phase quadrature of force noise
ρ	probability density
	in Chapters 8 to 10: density operator
J	coefficient of coupling between resonators
Δk	detuning spring force
U	in Chapter 6: normal-mode transformation matrix
	in Chapter 10: Kerr nonlinearity
ω_Δ	$= \frac{J}{\omega_0 m}$; angular exchange rate and spectral splitting
t_Δ	$= \frac{2\pi}{\omega_\Delta}$; energy exchange time
g	parametric coupling modulation depth
ω_R	angular Rabi frequency
$\hbar$	$\approx 1.05 \times 10^{-34}\,\mathrm{J\,s^{-1}}$; reduced Planck constant
σ_τ	state lifetime
σ_E	energy uncertainty
Ψ	wave function
n	in Chapters 8 to 10: Fock state number
a	$= \hat{a}$; annihilation operator
$a^\dagger$	$= \hat{a}^\dagger$; creation operator
x_{dl}	$= \frac{1}{2}(a^\dagger + a)$; dimensionless x operator
p_{dl}	$= \frac{i}{2}(a^\dagger - a)$; dimensionless x operator
α	amplitude of coherent state
P_j	probability of measuring the system in the state j
κ	$= \Gamma$; system-environment coupling rate
n_{th}	mean thermal excitation
U_{rot}	rotating-frame transformation matrix
Δ	$\omega - \omega_0$; angular frequency detuning
$\tilde{a}$	annihilation operator in the rotating frame
$\tilde{a}^\dagger$	creation operator in the rotating frame
α_R	real part of coherent state amplitude
α_I	imaginary part of coherent state amplitude
Δ_U	$= \Delta + U$; detuning shifted by the Kerr nonlinearity

Introduction

"It's still magic even if you know how it's done."
(Terry Pratchett, *A Hat Full of Sky*)

About This Work

This book emerged from a master-level course on "Parametric Phenomena" that the authors held together at ETH Zurich between 2018 and 2021, and individually at their respective universities since then. The course was organized as a reverse-classroom event: students would prepare by reading material at home, and then use the time in class to solve exercises and discuss with the teaching team. With this approach, we hoped to present the topic in much the same way as we experience it during our own research, and to encourage the students to formulate (and solve) their own questions. In line with this philosophy, the graded deliverable that every student handed in for passing the course was a poster that approximated one particular system as a parametric oscillator, including physical units and estimated numerical values. We saw many creative results, ranging from an airplane wobbling in the wind and a ship rolling in the sea to a nanoparticle trapped in an optical potential, a Josephson superconducting resonator, an optical ring resonator, a yo-yo, and even the predator–prey dynamics between a pack of wolves and a flock of sheep. This book is meant to provide all that is necessary to hold such a course, including the reading material, exercises, codes to solve the exercises, and a tutorial of how to map realistic physical systems onto the desired equations.

In this book, we perform a diagonal cut through many different topics. We follow a path from the deterministic mechanics of a harmonic oscillator all the way to the non-deterministic physics of coupled nonlinear quantum oscillators. Along this trajectory, we encounter many ideas and concepts that can fill entire books of their own accord. Our discussions of these concepts are guided by the wish to build an understanding without dealing with all possible details. This book is clearly *not* an exhaustive resource on topics such as nonlinear mechanics, stochastic physics, or the quantum oscillator. These topics have been treated in much more detail in other articles and books which we cite where appropriate. Rather, we want to focus on the combination of all these fundamental theories to gain a balanced and comprehensive view of the parametric oscillator.

In Chapter 1, we start with the deterministic behavior of the classical harmonic oscillator subject to damping and driving, and later to parametric pumping. Building on this foundation, we add nonlinearities in Chapter 2, and combine them with a parametric pump in Chapter 3. In Chapter 4, we introduce fluctuating forces for the example of the harmonic oscillator, which we generalize to the nonlinear parametric oscillator in Chapter 5. Coupling between oscillators is discussed in Chapter 6 and applied to stochastic, nonlinear parametric oscillators in Chapter 7. The quantum harmonic oscillator follows in Chapter 8, which leads to the driven and damped quantum harmonic oscillator in Chapter 9, and the quantum parametric oscillators in Chapter 10. Finally, in Chapter 11 we explain with several examples how mechanical, electrical, and optical systems can all serve as parametric oscillators.

0.1 Historical Review

In this Introduction, we review historical examples of parametric phenomena and understand why this topic is still the focus of so many research fields today. Before we can embark on our tour through the centuries, we must clarify what we mean by the term *parametric*. In our usage of the word, it refers to a periodic modulation of a resonator's potential — physically, the modulation could originate from a change in the tension of a mechanical string, a child alternatively standing and squatting on a swing, the effect of waves hitting a ship to change its buoyancy center, a variation in the effective capacitance of an electrical resonator, or an increase of the polarization of an optical medium in response to electromagnetic waves. All of these seemingly disparate examples obey very similar equations, and many of them can be used for similar technological applications (although so far no applications have been developed for children on swings . . .). The phenomena that arise as a consequence of parametric modulation are as varied as the physical systems in which they appear. At first glance, the menagerie of parametric phenomena may appear endless, but we will see that they all follow a few intuitive rules and can be classified accordingly.

The earliest examples of parametric oscillation are found, not surprisingly, in the mechanical domain. To our knowledge, the first experimental description of parametric resonance is ascribed to the works of Michael Faraday in 1831 [1] and Franz Melde in 1860 [2]. However, applications of the effect are much older: the big censer "O Botafumeiro" used for certain rituals in the Cathedral of Santiago de Compostela in Spain is set into pendulum motion by periodically modulating (i.e. parametrically pumping) the length of its rope [3]. As the censer weights about 60 kg and moves 20 m up and down during its largest oscillations, a team of operators is needed for this pumping, and their actions have to be coordinated in time to achieve the desired effect. Reports of parametric pumping of O Botafumeiro reach back to the 13th century. A mathematical treatment of parametric oscillation was not attempted until 1883, when Lord Rayleigh published his paper "On maintained vibrations" [4]. He analyzed the different types of driving that a system can experience and showed that parametric modulations can explain Faraday's experimental observations [1].

Technological applications of parametric pumping in electronics began to appear in the 20th century with the development of the Mag Amp [5] and the Klystron [6] amplifiers, both of which were based on time-dependent modulation of a control parameter. The Mag Amp found application in early radio telephones around 1915, and the Klystron allows high-power microwave generation and is still in use today for niche applications such as spacecraft communication and synchrotrons. In the second half of the century, inventions like the *Parametric Amplifier* by Arthur Ashkin and colleagues in 1959 [7] and the *Broadband cavity parametric amplifier with tuning* by Closson in 1962 [8] opened up new perspectives for electrical signal amplification. It was understood that a modulation of the reactance (i.e., the capacitance or inductance) of a resonant electrical circuit can lead to strong signal amplification without adding Nyquist noise which is unavoidable in resistor-based operational amplifiers [9, 10].

With the advent of superconducting circuits and the possibility of a strong nonlinear inductance imposed by Josephson junctions, the parametric amplifier was brought to its logical culmination, offering signal amplification with no more noise than what is absolutely required by the laws of quantum mechanics [11–13]. However, it was only after the turn of the millennium that these *Josephson parametric amplifiers* moved fully into the focus of the experimental quantum physics community [14–20], enabling experiments that previously were unfeasible [21]. Around the same time, parametric amplification [22–27] and coupling [28–32] were also explored in the growing nanomechanics community. A particularly important application arose in *cavity optomechanics*, where the parametric coupling between a mechanical and an optical degree of freedom can be used for precise control of the resonator and for cooling it down to its quantum ground state [33]. Parametric squeezing can be used to reduce fluctuations [22, 34–36] and has been employed as a means to generate nonclassical optical [37, 38] or mechanical [39–41] states. Importantly, parametric squeezing has been proposed as a way to boost the sensitivity of optical interferometers for gravitational wave detection [42–44].

Most of the above applications are achieved for relatively weak parametric modulation. By contrast, when the pumping exceeds a certain threshold, entirely new phenomena appear. Under strong parametric pumping at a frequency close to twice its resonance frequency, a resonator experiences a negative effective damping, such that it will ring up to large amplitudes and be stabilized only by nonlinear potential terms [45]. Such parametric instability appears in many contexts; for instance, it is held responsible for the dreaded *parametric rolling* of ships that has caused catastrophic accidents [46]. It is also considered as a possible mechanism for particle creation in models of the early universe [47, 48].

Beyond the instability threshold, the parametrically driven resonator can select one of two oscillation phases that are separated by a phase of π. This causes a spontaneous breaking of the time-translation symmetry of the system — oscillations with either phase are equivalent solutions in response to the drive, but only one of them can be realized at the same time (in a classical system) [45, 49]. Around 1960, Eiichi Goto [50] and John von Neumann [51] independently realized that these phase states offer a way to encode digital information. The *parametron* was indeed used as a memory unit for electrical computers in Japan until the invention of the transistor provided a more efficient solution.

0.2 Present and Future

Over the last few years, the development of novel resonators in the electrical, mechanical and optical domain has led to a revival of interest in the parametric oscillator[1] and the idea of *parametron phase logic*, in both the classical and quantum domains [52–65]. Of particular interest is the idea of coupling many parametrons into a configurable

[1] Other terms for the parametric oscillator are *Kerr parametric oscillator* or *two-photon driven Kerr resonator*.

Hopfield-type network [66, 67]. Here, the phase states of a single parametron represent the two polarization states of a spin, and the entire network can be used to simulate the behavior of the corresponding many-body Ising model [68]. Many optimization problems, such as the MAX-CUT problem [69, 70] or the number partitioning problem [71], are isomorphic to finding the ground state of an Ising network, and at the same time are nearly intractable with classical (sequential) computers [72]. Recent years have therefore seen a surge of ideas related to parametron logic control [55, 57, 73–77] and parametric network operation [62, 69–71, 78–86].

Whether the complexity of a multimode nonlinear oscillator network can be tamed to enable parallel computing and quantum simulations is an open question and will be the subject of intensive research over the coming years. What is safe to predict is that every new physical implementation of the harmonic oscillator sooner or later rediscovers parametric phenomena and applies it to a new purpose. A concept that is so versatile and useful will remain important in science and technology, no matter what the future brings.

1

The Harmonic Resonator

Harmonic oscillators are ubiquitous in nature and have been treated in many textbooks in depth [87, 88]. We briefly repeat in this first chapter those features that are important for the rest of the book. To facilitate an intuitive approach, we adopt the language of a mechanical oscillator, but the discussion may easily be translated to any oscillating system, cf. Chapter 11. Examples will be calculated without units, to preserve the spirit of a general treatment.

1.1 Newton's Equation of Motion

Consider a mass on a spring, see Fig. 1.1. The system has kinetic and potential energy, where the latter is stored in the spring proportional to the square of the displacement x, such that

$$H = E_{\text{tot}} = E_{\text{kin}} + E_{\text{pot}} = \frac{p^2}{2m} + \frac{1}{2}kx^2 \,. \tag{1.1}$$

Here, H is the Hamiltonian of the system, p the momentum and canonical conjugate of the displacement x, k the spring constant, and m the mass. The Hamiltonian is a function that describes the total kinetic and potential energy of a closed system. From Hamiltonian mechanics, we can calculate the force that acts on the mass at any given time t as

$$F \equiv \dot{p} \equiv \frac{dp}{dt} = -\frac{\partial H}{\partial x} = -kx \,, \tag{1.2}$$

where dots denote differentiation with respect to time t. The quadratic potential, thus, corresponds to a linear spring force. Combining eqn (1.1) with the second one of Hamilton's equations of motion (EOM),

$$\dot{x} = \frac{\partial H}{\partial p} = \frac{p}{m} \,, \tag{1.3}$$

we obtain a second-order differential equation that is known as Newton's EOM,

$$\ddot{x} + \frac{k}{m}x = 0 \,. \tag{1.4}$$

Equation (1.4) is solved using the ansatz $x(t) = x_{\text{ini}}e^{i\omega_0 t}$, where x_{ini} is determined by the initial boundary conditions, $\omega_0 = (k/m)^{1/2} = 2\pi\nu_0 = 2\pi/T_0$ is the angular resonance frequency, T_0 is the unforced periodicity of the oscillator, and we refer to

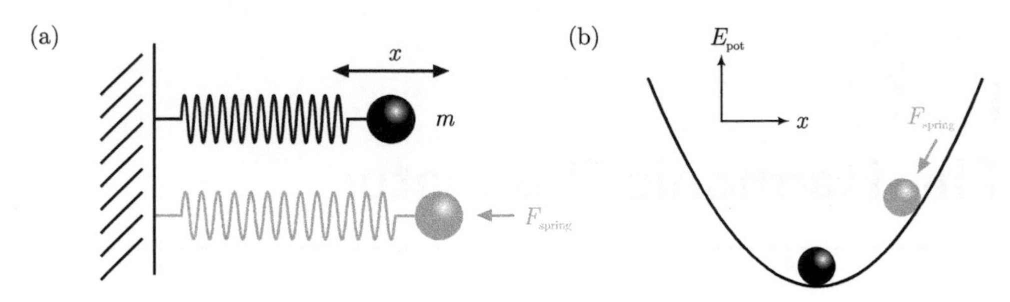

Fig. 1.1 (a) As an example of a harmonic oscillator, we use a mass on a spring. Displacing the mass from its rest position by x results in a restoring spring force $F_{\text{spring}} = -kx$. A displaced mass is shown in gray. (b) The potential energy of a harmonic oscillator is quadratic in displacement, $E_{\text{pot}} = \frac{1}{2}kx^2$, cf. eqn (1.2).

ν_0 as *natural frequency*. Note that eqn (1.4) describes an oscillator that is isolated from its environment, that is, Hamiltonian evolution is energy-conserving and does not feature damping terms.

Finding the microscopic origin of damping terms is an important topic on its own [89]. For now, we assume a phenomenological source of dissipation that enters Newton's EOM and can stabilize the oscillator's motion,

$$\ddot{x} + \Gamma\dot{x} + \frac{k}{m}x = 0\,, \tag{1.5}$$

where Γ is the coefficient corresponding to the dissipative (linear) damping enacted by the environment. Note that from a mathematical point of view, we can account for the added damping term through the transformation [88]

$$x(t) = e^{-\Gamma t/2}y(t) = e^{-t/\tau_0}y(t)\,, \tag{1.6}$$

where we define a decay time $\tau_0 = 2/\Gamma$. The equation of motion for $y(t)$ then takes the form of a closed harmonic oscillator,

$$\ddot{y} + \omega_\Gamma^2 y = 0\,, \tag{1.7}$$

in an exponentially expanding or shrinking coordinate system and with a slightly shifted resonance frequency

$$\omega_\Gamma^2 = \omega_0^2 - \frac{\Gamma^2}{4}\,. \tag{1.8}$$

From the transformation in eqn (1.6), we observe that for $2\omega_0 > \Gamma > 0$ the oscillator coordinate $x(t)$ decays exponentially in time in addition to an harmonic oscillation. However, we can already guess that something different must happen once $2\omega_0 \leq \Gamma$.

A direct treatment of the homogeneous dissipative case in eqn (1.5) is possible starting from the same ansatz that any particular solution has the form

$$x(t) = x_{\text{ini}}e^{\mu t} \tag{1.9}$$

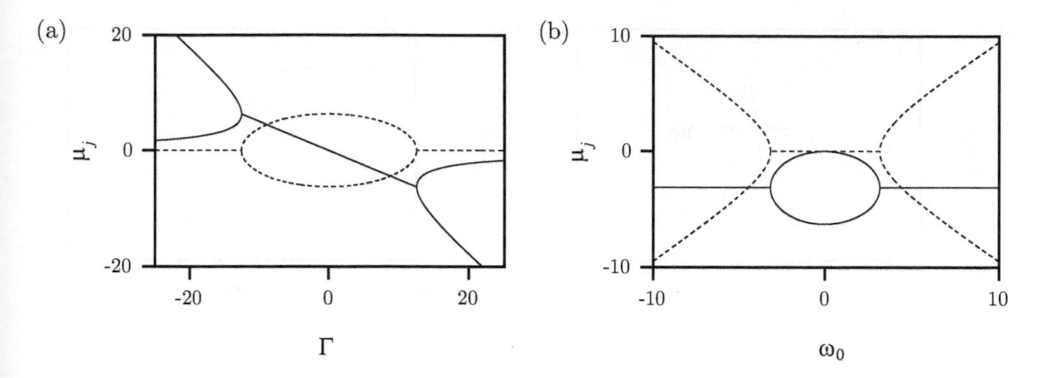

Fig. 1.2 The real (solid) and imaginary (dashed) parts of the characteristic exponents, cf. eqn (1.11), as a function of (a) damping coefficient Γ for a bare angular resonance frequency $\omega_0 = 2\pi$, and of (b) ω_0 for $\Gamma = 2\pi$.

with a complex **characteristic exponent** $\mu \in$. Inserting eqn (1.9) into eqn (1.5) leads to

$$x\left(\mu^2 + \mu\Gamma + \omega_0^2\right) = 0\,, \tag{1.10}$$

which, for $x \neq 0$, results in a quadratic **characteristic equation** with the two roots

$$\mu_{a,b} = -\frac{\Gamma}{2} \pm \sqrt{\Gamma^2/4 - \omega_0^2} = -\frac{\Gamma}{2} \pm i\omega_\Gamma\,. \tag{1.11}$$

This is identical to what we obtained with the coordinate transformation method in eqn (1.6), see eqn (1.8) and the discussion thereafter.

We can identify several distinct regimes of motion: for damped oscillators ($\Gamma > 0$) we distinguish between overdamped ($\omega_\Gamma^2 < 0$), critically damped ($\omega_\Gamma^2 = 0$), and underdamped motion ($\omega_\Gamma^2 > 0$), where oscillation appears only for the latter.[1] For $\Gamma < 0$, the oscillator is unstable and the motion becomes unbounded. This is visualized by plotting the real and imaginary part of the characteristic exponents $\mu_{a,b}$, see Fig. 1.2. Note that, in many cases, the small correction to the bare frequency due to the damping term is neglected, such that $\omega_\Gamma^2 \approx \omega_0^2$.

1.2 Response of the Driven Resonator

In large parts of our treatment, we will use Newton's EOM to analyze the behavior of driven oscillating systems. For our mass on a spring, we can write

$$\ddot{x} + \frac{k}{m}x + \Gamma\dot{x} = \frac{F_0}{m}\cos(\omega t)\,, \tag{1.12}$$

where $F = F_0\cos(\omega t)$ is an external driving force that turns eqn (1.12) into an inhomogeneous differential equation.

[1] The critically damped point is an example of an *exceptional point* where the roots are degenerate and eqn (1.9) is insufficient [90].

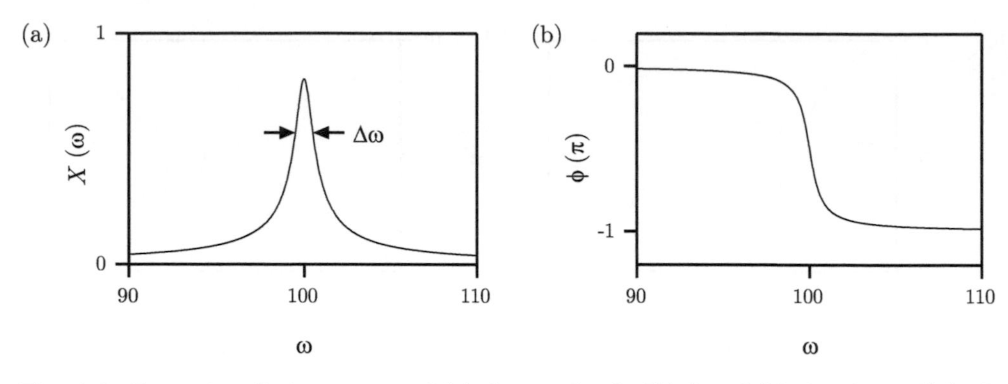

Fig. 1.3 Long-time limit response of (a) the amplitude $X(\omega)$ and (b) the phase $\phi(\omega)$ of a driven resonator, cf. eqns (1.14) and (1.15), respectively. $F_0/m = 80$, $\omega_0 = 100$ and $\Gamma = 1$ ($Q = 100$).

For a driven, damped resonator, we proceed by Fourier transforming eqn (1.12). The Fourier component of F at ω is simply F_0. Solving for the frequency-dependent, complex amplitude $x(\omega) = X(\omega)e^{i\phi(\omega)}$ yields

$$x(\omega) = \frac{F_0/m}{\omega_0^2 - \omega^2 + i\omega\Gamma} \equiv F_0\chi\,, \tag{1.13}$$

where we use χ to denote the susceptibility of the resonator. Note that $x(\omega)$ is a complex number, while $X(\omega)$ and $\phi(\omega)$ are assumed to be real. In many cases, we are interested in the long-time limit response of the resonator to a force at a single frequency, and find

$$|x(\omega)| = X(\omega) = \frac{F_0/m}{\sqrt{\left(\omega_0^2 - \omega^2\right)^2 + \omega^2\Gamma^2}}\,, \tag{1.14}$$

as well as

$$\phi(\omega) = \tan^{-1}\left(\frac{\mathrm{Im}[x(\omega)]}{\mathrm{Re}[x(\omega)]}\right) = \tan^{-1}\left(\frac{\omega\Gamma}{\omega_0^2 - \omega^2}\right)\,. \tag{1.15}$$

The amplitude and phase response functions are drawn in Fig. 1.3. Experimentally, such curves are often used to extract the resonance frequency and quality factor of a resonator: the full-width-at-half-maximum (FWHM) of the **squared amplitude** $X^2(\omega)$ is given by $\Delta\omega = \Gamma = \omega_0/Q$, with Q the quality factor (or, if temporal frequency is plotted, $\Delta\nu = \nu_0/Q$). If the unsquared response is measured, such as in Fig. 1.3(a), one must extract $\Delta\omega$ or $\Delta\nu$ at $\sqrt{1/2}$ of the maximum.

1.3 Matrix Formulation

In this section, we will introduce the technique of the state transition matrix method [88]. Generally, any N-order ODE can be cast into a set of N coupled first-order ODEs. We will here cover the case of a second-order differential equation. We start by casting eqn (1.12) into two coupled first-order differential equations (cf. Hamilton's EOMs above), which are written in matrix form as

$$\dot{\mathbf{x}}(t) = G(t)\mathbf{x}(t) + \mathbf{f}(t)\,, \tag{1.16}$$

where

$$\mathbf{x}(t) = \begin{bmatrix} x(t) \\ \dot{x}(t) \end{bmatrix} \tag{1.17}$$

is a state vector describing the resonator coordinates, $G(t)$ is a matrix that contains the coefficients, and $\mathbf{f}(t)$ is the forcing vector. In the concrete example of eqn (1.12), we write

$$\begin{bmatrix} \dot{x}(t) \\ \ddot{x}(t) \end{bmatrix} = \begin{bmatrix} 0 & 1 \\ -\omega_0^2 & -\Gamma \end{bmatrix} \begin{bmatrix} x(t) \\ \dot{x}(t) \end{bmatrix} + \begin{bmatrix} 0 \\ F_0 \cos(\omega t)/m \end{bmatrix}. \tag{1.18}$$

1.3.1 Solving the Equation in Matrix Form

We can use eqn (1.16) for an alternative derivation of the solution of the second-order differential equation shown in eqn (1.11) [88, 91, 92]. Namely, for $\mathbf{f}(t) = 0$ and assuming $x(t)$ to be a particular solution of the general form $x_{\mathrm{ini}}e^{\mu t}$, we know that $\dot{\mathbf{x}}(t) = \mu\mathbf{x}(t)$, therefore we obtain in combination with eqn (1.16)

$$\dot{\mathbf{x}}(t) = G\mathbf{x}(t) = \begin{bmatrix} \mu & 0 \\ 0 & \mu \end{bmatrix} \mathbf{x}(t)\,. \tag{1.19}$$

For $x_{\mathrm{ini}} \neq 0$, this equation is only fulfilled if

$$\left| G - \begin{bmatrix} \mu & 0 \\ 0 & \mu \end{bmatrix} \right| = \mu^2 - \mathrm{tr}(G)\mu + |G| = 0\,, \tag{1.20}$$

where $|...|$ stands for the determinant of a matrix and tr(...) for its trace. Equation (1.20) is another way to write the **characteristic equation** which we have already seen in eqn (1.10) for the case of a dissipative harmonic oscillator. From the result of the characteristic equation, we can determine whether the system decays in time ($\mu < 0$) or grows without bound ($\mu > 0$). Note also that eqn (1.20) is equivalent to solving for the eigenvalues of G.

1.3.2 Stability Analysis

Equation (1.16) offers a simple way to find the stationary points of the system by imposing $\dot{\mathbf{x}}(t) = 0$. In the case above with $\mathbf{f}(t) = 0$, we obtain $\mathbf{x} = 0$ as the only stationary point in the system. The dynamics around this point are then described by eqn (1.20).

This procedure can be similarly performed for other stationary points. Specifically, having found an equilibrium point, we ask ourselves if the point is stable against small perturbations in $\mathbf{x}$; does the system return to the equilibrium point after it has been slightly displaced? There is a simple method to answer this question: assuming

$(x_{\rm eq}, \dot{x}_{\rm eq})$ to be an equilibrium point where $\dot{\mathbf{x}}(t) = 0$, we Taylor expand to linear order around this point. Our new degrees of freedom are now

$$\delta\mathbf{x}(t) = \begin{bmatrix} \delta x(t) \\ \delta\dot{x}(t) \end{bmatrix} = \begin{bmatrix} x(t) \\ \dot{x}(t) \end{bmatrix} - \begin{bmatrix} x_{\rm eq}(t) \\ \dot{x}_{\rm eq}(t) \end{bmatrix} . \tag{1.21}$$

Similar to eqn (1.19) and eqn (1.20), we can formulate a characteristic equation for $\delta\mathbf{x}$ to find out if the system grows away from the equilibrium point $(x_{\rm eq}, \dot{x}_{\rm eq})$ over time (unstable behavior) or will decay toward it (stable). A point is only stable if all real characteristic exponents are negative. The underdamped oscillator in a quadratic potential is only stable at $\mathbf{x} = 0$.

1.3.3 From a Differential to an Integral Equation

The matrix formulation is useful not only for analysis of the system's equilibrium points and their stability to small perturbations, but also for calculating the time evolution given any specific initial conditions and external force [88]. In order to show this, we first define the Wronskian matrix

$$W(t) = \begin{bmatrix} x_a(t) & x_b(t) \\ \dot{x}_a(t) & \dot{x}_b(t) \end{bmatrix} . \tag{1.22}$$

The entries of the Wronskian matrix in the first row, $x_a(t)$ and $x_b(t)$, are orthogonal basis solutions to the homogeneous EOM. In our case of eqn (1.5), we can choose for example $x_a(t) = \cos(\omega_0 t)e^{-\Gamma t/2}$ and $x_b(t) = \sin(\omega_0 t)e^{-\Gamma t/2}$. The determinant of the Wronskian matrix, which must be nonzero in order to make the matrix invertible, is simply called the **Wronskian**. For the homogeneous case, a general solution is of the form $\mathbf{x}(t) = W(t)\mathbf{d}$ where $\mathbf{d}$ is a vector of constant coefficients to be defined through the boundary conditions. Moving to an inhomogeneous case by applying an external drive $\mathbf{f} \neq 0$, we use a so-called *variational ansatz* that allows the prefactors to also depend on time. This leads to

$$\mathbf{x}(t) = W(t)\mathbf{d}(t) . \tag{1.23}$$

Differentiating this solution, we obtain

$$\dot{\mathbf{x}}(t) = \dot{W}(t)\mathbf{d}(t) + W(t)\dot{\mathbf{d}}(t) , \tag{1.24}$$

where the first term on the right-hand side satisfies

$$\dot{W}(t)\mathbf{d}(t) = G(t)W(t)\mathbf{d}(t) \tag{1.25}$$

because $W(t)$ is, by definition, composed out of particular solutions of eqn (1.16) for $\mathbf{f}(t) = 0$. Therefore, if we enter eqn (1.23) into eqn (1.16), we obtain

$$W(t)\dot{\mathbf{d}}(t) = \mathbf{f}(t) . \tag{1.26}$$

Multiplying this equation with $W^{-1}(t)$ from the left and integrating $\dot{\mathbf{d}}(t)$ over time, we find that

$$\mathbf{d}(t) = \mathbf{d}(0) + \int_0^t W^{-1}(t')\mathbf{f}(t')dt' \tag{1.27}$$

with $\mathbf{d}(0) = W^{-1}(0)\mathbf{x}(0)$ and t' denoting the integration variable. The time evolution of $\mathbf{x}(t)$ then follows from eqn (1.23) as

$$\mathbf{x}(t) = \Phi(t,0)\mathbf{x}(0) + \int_0^t \Phi(t,t')\mathbf{f}(t')dt' \,, \tag{1.28}$$

where we have introduced the state transition matrix

$$\Phi(t,t') = W(t)W^{-1}(t') \,. \tag{1.29}$$

For the concrete case of basis solutions chosen as $x_a(t) = \cos(\omega_0 t)e^{-\Gamma t/2}$ and $x_b(t) = \sin(\omega_0 t)e^{-\Gamma t/2}$, we obtain

$$\Phi(t,t') = e^{\frac{1}{2}\Gamma(t'-t)} \begin{bmatrix} \frac{(\Gamma \sin(\omega_0(t-t'))+2\omega_0 \cos(\omega_0(t-t')))}{2\omega_0} & \frac{\sin(\omega_0(t-t'))}{\omega_0} \\ -\frac{(\Gamma^2+4\omega_0^2)\sin(\omega_0(t-t'))}{4\omega_0} & \frac{(2\omega_0 \cos(\omega_0(t-t'))-\Gamma \sin(\omega_0(t-t')))}{2\omega_0} \end{bmatrix} . \tag{1.30}$$

This formalism produces the same long-time limit solutions as eqn (1.13), but furthermore allows us to study time-dependent phenomena. We understand from eqn (1.26) that the role of the force term $\mathbf{f}(t)$ is to change the entries of the weighting vector $\mathbf{d}$ over time; for instance, a force may increase the amplitude of oscillation, which will appear as a larger entry in $\mathbf{d}(t)$. The force can also change the phase of an oscillation, such that it evolves from $x_a(t) = \cos(\omega_0 t)e^{-\Gamma t/2}$ to $x_b(t) = \sin(\omega_0 t)e^{-\Gamma t/2}$. In addition, the forcing can lock the motion of the oscillator to an external driving frequency. In the absence of an external force, the evolution is entirely determined by the initial condition, which corresponds to $\mathbf{d}(0)$.

Equation (1.28) implies that when the fundamental solutions basis and the initial conditions are known, the complete time evolution of a linear differential equation can be analytically determined. Importantly, this is even possible in cases where the entries of $G(t)$ vary in time, as long as fundamental solutions are known at any given moment. The integration in eqn (1.28) is then performed piecewise.

The result of a ringdown calculated with eqn (1.28) is shown in Fig. 1.4(a). Ringdown measurements are useful in experiments because they allow for a precise characterization of the damping. In Fig. 1.4(b), an example with external forcing is provided. The evolution depends crucially on the phase of the force relative to the starting conditions. However, the long-time limit response is independent on $\mathbf{x}(0)$ and corresponds to the result found in eqn (1.14).

1.4 Parametric Modulation

Our main focus will be on the study of parametric resonators, that is, systems where the homogeneous terms of the model, such as the potential energy coefficient (the spring constant), are allowed to vary periodically in time, see Fig. 1.5. In this section,

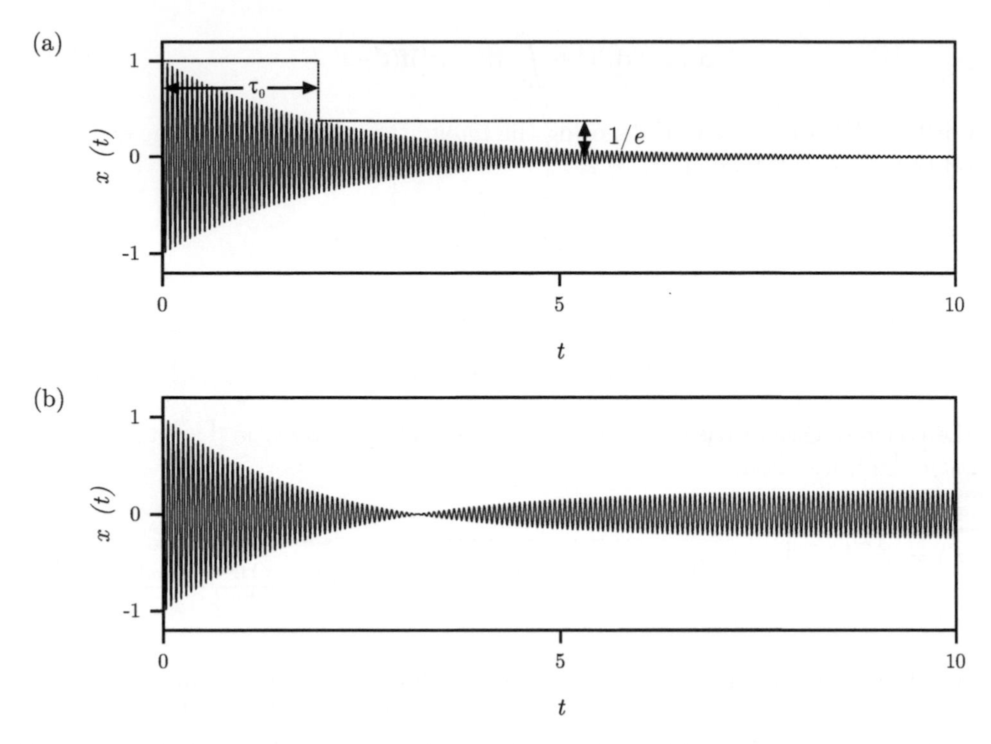

Fig. 1.4 (a) Ringdown timetrace of an undriven, damped harmonic resonator calculated with the state transition matrix method, cf. eqn (1.28), for $\omega_0 = 100$ and $\Gamma = 1$ ($Q = 100$). The amplitude decays exponentially as e^{-t/τ_0} with a decay time $\tau_0 = 2/\Gamma = 2Q/\omega_0$. (b) The same system evolving under the influence of an external force with $F_0/m = 25$. Due to the specific phase of the force, the system rings down faster than in the unforced case, and then rings up to a stable amplitude with an inverted phase relative to the starting condition.

we gain a first impression of the behavior of such systems. We start with the undamped **Hill equation**,

$$\ddot{x} + \omega_0^2 \left[1 + \lambda \Psi_H(t)\right] x = 0\,, \tag{1.31}$$

where $\Psi_H(t)$ is a function that is periodic over a time T_p, and where λ is a constant that characterizes the modulation depth. The most widely used form of the Hill equation is the **Mathieu equation**,

$$\ddot{x} + \omega_0^2 \left[1 + \lambda \cos(\omega_p t + \psi)\right] x = 0\,, \tag{1.32}$$

with $\omega_p \equiv 2\pi/T_p$ and a constant phase ψ. The same transformation as in eqn (1.6) can be applied to the Hill equation to analyze damped systems. In later chapters, however, we will treat damping explicitly to facilitate observation of the systems in a *lab frame*.

The effect of resonant parametric pumping at $\omega_p \approx 2\omega_0$ is fundamentally different from that of external forcing because the pump term is linear in x. On the one hand, starting from the initial condition $x = \dot{x} = 0$ the system will undergo no evolution at all. On the other hand, if the system has nonzero initial conditions and is oscillating

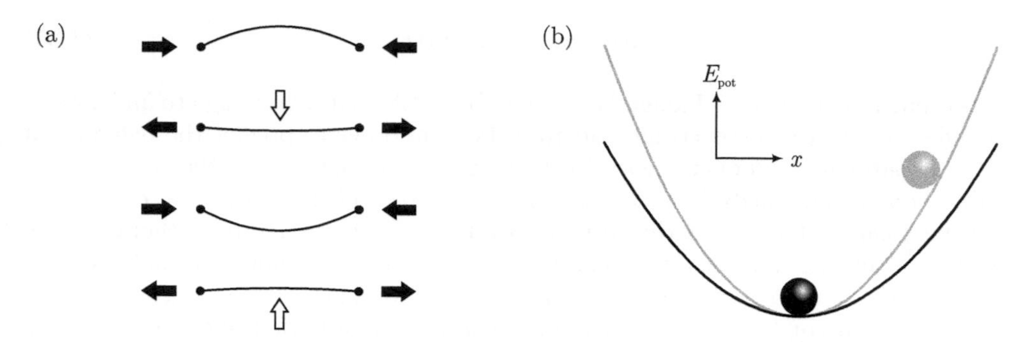

Fig. 1.5 (a) A string is held at two clamping points (round dots). The four pictures show snapshots in time. Solid arrows indicate positive and negative tension that is applied to the string at the clamping points, while hollow arrows correspond to the instantaneous lateral string velocity. The modulated tension is a form of parametric pumping and can amplify or damp the motion depending on the relative phase. (b) Parametric pumping corresponds to a modulation of the harmonic potential shape as a function of time, as indicated by the two different snapshots in black and gray.

as $x(t) \propto \cos(\omega_0 t)$, the multiplication of $x(t)$ with $\lambda \cos(\omega_p t)$ will generate an effective driving term $\propto \cos((\omega_p - \omega_0)t)$. Since it is proportional to the instantaneous oscillation amplitude, this parametric pumping will act as an additional positive or negative damping term Γ_p for different relative phases ψ between the vibration and the pump. If the resonator has no internal damping, that is $\Gamma = 0$, the effective damping $\Gamma_{\text{eff}} = \Gamma + \Gamma_p$ will become negative for two opposite phases of motion (two values of the phase that are separated by π). The resonator will therefore experience exponential amplification in these phases and will become unstable. In the following, we will demonstrate this behavior using Floquet analysis.

1.5 Floquet Theory

When the coefficients of the homogeneous system are subject to periodic modulations with a period T_p, the system may become unstable even in the presence of positive damping ($\Gamma > 0$) [88]. In order to investigate the conditions for stability of a modulated system, we first note that the state transition matrix Φ fulfills the homogeneous part of eqn (1.16), that is,

$$\dot{\Phi}(t, 0) = G(t)\Phi(t, 0)\,. \tag{1.33}$$

From this, and taking into account that $G(t) = G(t+T_p)$, it follows that the operation that $\Phi(t, 0)$ performs on an initial state $\mathbf{x}(0)$ can be split into repeated identical parts that describe the operation over one modulation period T_p,

$$\Phi(T_p, 0) = C\,. \tag{1.34}$$

In the absence of forcing, the state evolution over n periods can then simply be described by

$$\mathbf{x}(t + nT_p) = C^n \mathbf{x}(t)\,. \tag{1.35}$$

This is one form to express Floquet's theorem. It entails that it is enough to understand the effect of the **discrete transition matrix** C in order to predict the behavior of $x(t)$. In particular, Floquet theory is often applied to analyze whether a system is stable or not. The analysis makes use of the complex eigenvalues $c = e^{\hat{\mu}T_p}$ of C, which are called **characteristic multipliers**. Here, $\hat{\mu}$ are referred to as characteristic exponents, but we have marked them with a hat symbol to emphasize that they are not identical to the characteristic exponents μ that we have previously encountered, for example in eqn (1.11). Indeed, C only contains information about the discrete changes that occur over full periods T_p, and not about the periodic evolution of $\mathbf{x}(t)$ within these time steps.

For the familiar example of the underdamped harmonic resonator, we use the natural periodicity of $T_p = 2\pi/\omega_\Gamma$. We then find that the two eigenvalues $c_{a,b}$ are real and equal and amount to the exponential decay factor $e^{-\Gamma T_p/2}$, such that $\hat{\mu} = -\Gamma/2$. By contrast, if at least one of the characteristic multipliers has a real component larger than 1, the system will grow infinitely. In reality, such growth is always bounded by the emergence of nonlinearities.

In the light of eqn (1.33), we write down a characteristic equation similar to eqn (1.20) to find the eigenvalues of $\Phi(T_p, 0) = C$. Here, we obtain

$$c^2 - \mathrm{tr}(\Phi(T_p, 0))c + |\Phi(T_p, 0)| = 0\,. \tag{1.36}$$

We can further use the relationships [88]

$$c_a c_b = \exp\left[\int_0^{T_p} \mathrm{tr}(G)dt\right] = |\Phi(T_p, 0)|\,. \tag{1.37}$$

As a concrete example, we will analyze the stability of the so-called Meissner equation with added damping,

$$\ddot{x} + \omega_0^2\,[1 - \lambda\psi_M(t)]\,x + \Gamma\dot{x} = 0\,, \tag{1.38}$$

where $\psi_M(t) = -1$ for $0 < t < T_p/2$ and $\psi_M(t) = 1$ for $T_p/2 < t < T_p$ within each period T_p. The Meissner equation has similar properties as the Mathieu equation and can be solved analytically by defining state transition matrices over partial periods. Namely, if the coefficients of an EOM vary over time, the discrete transition matrix C can be obtained from the multiplication of state transition matrices covering sub-period time intervals,

$$C = \Phi(T_p, 0) = \prod_{n=1}^{N} \Phi(t_n, t_{n-1})\,, \tag{1.39}$$

where $t_0 = 0$ and $t_N = T_p$. Note that for the Mathieu equation, this procedure would require infinitely many products because the potential term changes continuously.

We start with eqn (1.6) to transform our system from eqn (1.38) into

$$\ddot{y} + \omega_0^2 \left[1 - \lambda \psi_M(t)\right] y = 0 \,, \tag{1.40}$$

where we assume $Q \gg 1$ and therefore $\omega_0 \approx \omega_\Gamma$. The Meissner equation allows us to divide the full period T_p into two half periods with constant coefficients,

$$\ddot{y} + \omega_M^2 y = 0 \tag{1.41}$$

with

$$\begin{aligned} \omega_M &= \omega_0 \sqrt{1-\lambda} \qquad \text{for} \quad 0 \le t < T_p/2 \,, \\ \omega_M &= \omega_0 \sqrt{1+\lambda} \qquad \text{for} \quad T_p/2 \le t < T_p \,. \end{aligned} \tag{1.42}$$

For each half period, we can define fundamental solutions $\cos\left(\omega_M t\right)$ and $\sin\left(\omega_M t\right)$ and corresponding Wronskian matrices

$$W(t) = \begin{bmatrix} \cos\left(\omega_M t\right) & \sin\left(\omega_M t\right) \\ -\omega_M \sin\left(\omega_M t\right) & \omega_M \cos\left(\omega_M t\right) \end{bmatrix} , \tag{1.43}$$

which lead us to half-period state transition matrices

$$\Phi(t, t') = W(t) W^{-1}(t') \,. \tag{1.44}$$

The final state transition matrix is obtained from

$$\Phi(T_p, 0) = \Phi(T_p, T_p/2)\Phi(T_p/2, 0) \,, \tag{1.45}$$

where we remember that the coefficients in the half-period state transition matrices assume different values, see eqn (1.42).

In order to calculate the eigenvalues $c_{a,b}$, we use the characteristic equation (1.36). For a second-order system without loss, $\mathrm{tr}(G(t)) = 0$, so we can insert $|\Phi(T_p, 0)| = 1$ and arrive at

$$c_{a,b} = \frac{\mathrm{tr}(\Phi(T_p, 0))}{2} \pm \sqrt{\left(\frac{\mathrm{tr}(\Phi(T_p, 0))}{2}\right)^2 - 1} \,. \tag{1.46}$$

Finally, we transform back into x-coordinates and find the characteristic exponents as

$$\hat{\mu}_{a,b} = \frac{\log_e\left(c_{a,b}\right)}{T_p} - \frac{\Gamma}{2} \,. \tag{1.47}$$

The characteristic exponent of the undamped system can also be calculated directly [88], yielding the expression

$$\hat{\mu}_{a,b} = \pm \frac{1}{T_p} \cosh^{-1}\left(\frac{\mathrm{tr}(\Phi(T_p, 0))}{2}\right) - \frac{\Gamma}{2} \,. \tag{1.48}$$

Equation (1.47) or (1.48) can be used to map the stability of the damped Meissner oscillator as a function of the modulation depth λ and the detuning of the modulation

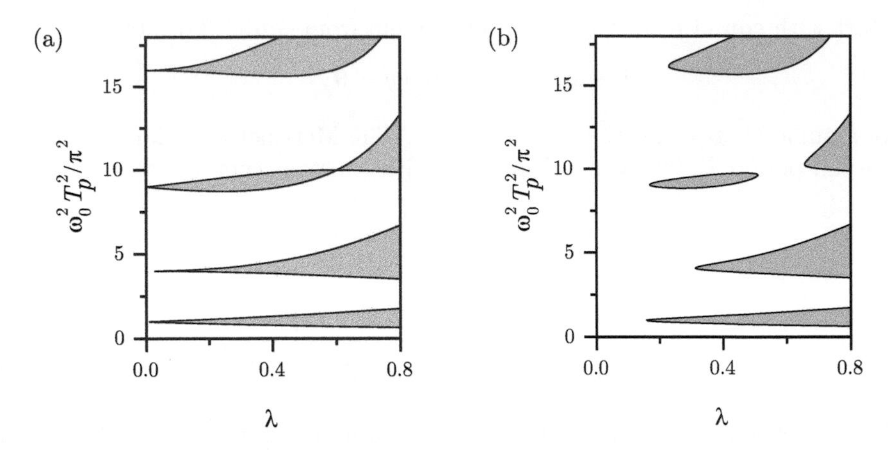

Fig. 1.6 (a) Stability diagram of the Meissner oscillator, cf. eqn (1.38), as a function of $\omega_0^2 T_p^2/\pi^2$ and λ. White areas correspond to a stable system with $\hat{\mu}_{a,b} < 0$, while gray areas correspond to an unstable system with $\hat{\mu}_{a,b} > 0$, cf. eqn (1.48). (b) The same system with added damping $\Gamma = 0.1$. The tongues are rounded off by the damping and no longer reach $\lambda = 0$.

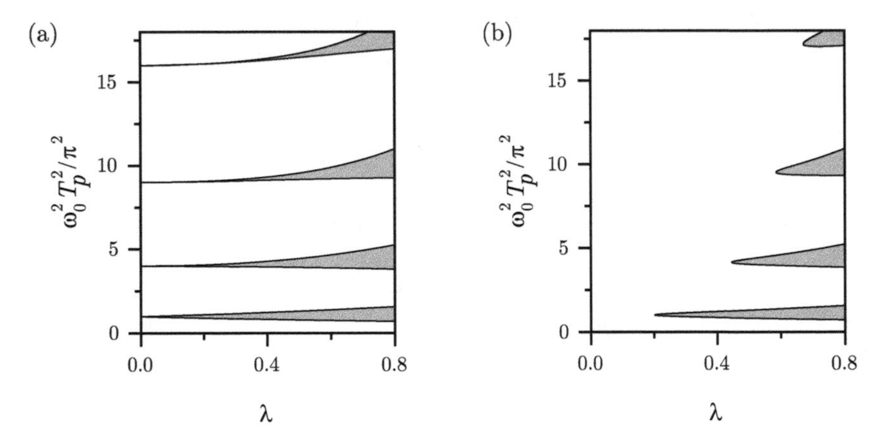

Fig. 1.7 (a) Stability diagram of the Mathieu equation, cf. eqn (1.32), as a function of $\omega_0^2 T_p^2/\pi^2 = 4\omega_0^2/\omega_p^2$ and λ. White areas correspond to a stable system with $\hat{\mu}_{a,b} < 0$, while gray areas correspond to an unstable system with $\hat{\mu}_{a,b} > 0$. (b) The same system with added damping $\Gamma = 0.1$. The tongues are rounded off by the damping and no longer reach $\lambda = 0$.

frequency $\omega_p = 2\pi/T_p$ from ω_0. The stability diagrams in Fig. 1.6 exhibit distinct instability lobes where the system diverges under the influence of the Meissner modulation. The gray regions are known as **Arnold's tongues**. For an undamped system, the Arnold's tongues stretch all the way to $\lambda = 0$, contacting the frequency axis at $\omega_0^2/\omega_p^2 = n^2/4$, where each natural number n indexes one tongue. This condition for the pump frequencies ω_p where instabilities occur is generally true for any Hill equation as long as they share the same Floquet period T_p. In Fig. 1.7, we can see the same set of tongues solved numerically for the Mathieu equation with $\omega_p = 2\pi/T_p$. In the following chapters, we will concentrate on the lowest Arnold tongue of the Mathieu equation where $\omega_p \approx 2\omega_0$.

Chapter summary

- In Chapter 1, we study the deterministic behavior of a driven and damped harmonic oscillator.
- The evolution of the system in the absence of damping or driving is described by **Hamilton's equations of motion**, cf. eqns (1.2) to (1.3). When damping or driving are added, we can instead use **Newton's equation of motion**, which is a linear second-order differential equation, cf. eqn (1.4).
- In the absence of driving, the equation of motion of the damped harmonic oscillator can be converted to that of an undamped harmonic oscillator by a coordinate transformation, cf. eqn (1.6). This step often helps to find simple solutions.
- As a function of the damping coefficient $\Gamma = \omega_0/Q$ relative to the angular resonance frequency ω_0, the damped harmonic oscillator undergoes a transition from an underdamped ($\omega_0 > \Gamma/2$) to an overdamped system ($\omega_0 < \Gamma/2$), cf. Fig. 1.2 and eqn (1.11). This relationship can be expressed in terms of the quality factor $Q \gtrless 1$. We generally concentrate on the regime of strongly underdamped resonators with $Q \gg 1$.
- In the presence of a near-resonant external force, an underdamped system is driven to oscillate. The amplitude and phase response in the long-time limit can be calculated with a single-frequency ansatz, cf. Fig. 1.3 and eqn (1.13).
- The response of the damped harmonic oscillator to an external force can be treated with a matrix formalism. On the one hand, this formalism offers an elegant way to determine the equilibrium points and their stability via the characteristic equation, cf. eqns (1.20) and (1.21). On the other hand, we can follow the full response in time after introducing the **Wronskian matrix**, cf eqn (1.22), and the **state transition matrix**, cf. eqn (1.28).
- The state transition matrix method facilitates the introduction of **Floquet's theory**. For the purpose of this book, we use the Floquet formalism to study the stability of a harmonic oscillator in the presence of parametric modulation. Specifically, we use the **Meissner equation** as a case study to demonstrate that parametric pumping can make a system unstable even in the presence of damping, cf. Fig. 1.6 and eqn (1.47) or (1.48).

Exercises

Check questions:

(a) What is the relationship between the potential energy and the force? What are the general conditions for harmonic oscillation?

(b) What are the various phenomena caused by the damping term?

(c) What is the meaning of the oscillator's Q? Can you find three methods to calculate it from a numerical experiment?

(d) What is the meaning of the Wronskian and what is it used for? Why does it contain two independent solutions x_a and x_b?

(e) What is the difference between the state transition matrix and C? In general, what information are we missing in Floquet theory as introduced in Section 1.5?

Tasks:

1.1 Familiarize yourself with the Jupyter Notebook **Python Example 1**.

1.2 Set $\lambda = 0$ ("**lam=0**") and $F_0 = 0$ ("**F0=0**") and run single timetraces with various values of ω_0, Q, and initial conditions. What do you see? Is it in agreement with your expectations? Compare your numerical result to an appropriate analytical method from this chapter.

1.3 Pay attention to the sampling time step ("**dt**"). What condition should it fullfil for a faithful simulation (measurement)? In general, what is a useful check one can perform to test that the numerical result is not influenced by the sampling time step?

1.4 Add an external force **F0** and test if you can predict the outcome of an experiment with the response function in eqn (1.13). Use $\omega_0 = 1$, $m = 1$, and $\Gamma = 0.01$ as starting values. Can you formulate a requirement for the final time **tf** to obtain good agreement between the analytical and numerical results?

1.5 Using the provided fast Fourier transform (FFT) function, study the frequency-domain results of various runs. How does the spectrum of a ringdown look compared to that of a driven resonator in the short-time or long-time limit? Can you extract Q from the spectrum in any of these cases?

1.6 Run the provided sweeper tool to simulate a "frequency sweep", where a single timetrace is performed at every frequency value of the sweep. Find two different methods to extract Q from the sweep result. Use $\omega_0 = 1$, $m = 1$, and $\Gamma = 0.01$ as starting values.

1.7 Set the external force back to $F_0 = 0$ and add a finite parametric modulation depth **lam** with $\omega = 2\omega_0$. Perform ringdowns with different parametric modulation depths and pumping phases **psi**. Why is there a phase dependency? Can you experimentally find above which modulation depth the oscillator becomes unstable?

1.8 Perform numerical ringdown scans for a parametric drive with frequency $\omega = 2\omega_0$. Keep λ small enough to avoid instabilities. Extract Q as a function of the phase and amplitude of the parametric drive and describe what you see. Can you propose a simple phenomenological model for this behavior?

2
The Duffing Resonator

So far, we have reviewed some properties of the harmonic resonator. We briefly showed that parametric pumping can lead to an unbounded exponential growth of the resonator amplitude if there is not enough damping. In reality, most physical systems will not actually explode in such a situation, but will eventually be stabilized by nonlinear terms. In this section, we familiarize ourselves with the most common nonlinearity, which is the Duffing term. For further reading, we recommend the following from the References: [45, 91, 92].

2.1 The Quartic Potential

In the same way that a quadratic potential energy $kx^2/2$ gives rise to a linear restoring force $-kx$, a quartic potential energy $\beta_3 x^4/4$ is responsible for a cubic restoring force term $-\beta_3 x^3$. This is the famous **Duffing nonlinearity**. Our Newton equation of motion is therefore modified as

$$\ddot{x} + \omega_0^2 x + \beta_3 x^3 + \Gamma\dot{x} = F_0 \cos(\omega t)/m. \tag{2.1}$$

As we will see, the inclusion of this innocuous term has far-reaching consequences. We will limit our discussion to the regime of small displacements, $X^2 \ll \omega_0^2/\beta_3$, where a negative Duffing nonlinearity, $\beta_3 < 0$, does not lead to an instability, see Fig. 2.1.

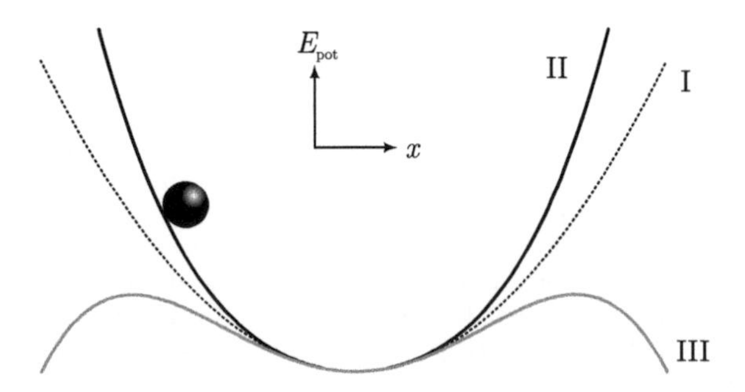

Fig. 2.1 A Duffing nonlinearity corresponds to a quartic potential energy term. Here, potential I has a purely harmonic shape, while II and III feature positive and negative Duffing terms, respectively. In the last case, the system becomes unstable for large displacement.

2.1.1 Four-Wave Mixing

In Section 1.3, we learned how a matrix formulation can be used to solve a differential equation. The central result was eqn (1.28), which offered a solution based on an initial condition $\mathbf{x}(0)$ and an external forcing $\mathbf{f}(t)$. We integrated eqn (1.28) over time to follow the evolution of $\mathbf{x}(t)$.

When introducing nonlinear terms, we encounter a problem with eqn (1.28). We want to solve the EOM

$$\ddot{x} + \omega_0^2 x + F(x, \dot{x}, t) = 0\,, \tag{2.2}$$

where we have for convenience defined the function

$$F(x, \dot{x}, t) = \beta_3 x^3 + \Gamma\dot{x} - F_0 \cos(\omega t + \theta)/m\,. \tag{2.3}$$

Note that $F(x, \dot{x}, t)$ also contains the damping term $\Gamma\dot{x}$, whose effect we previously handled as part of the orthogonal basis solutions described by the Wronskian, cf. eqn (1.22). In the following, we want to treat $\Gamma\dot{x}$ on an equal footing with nonlinear terms to make sure we do not overlook an important effect.

The additional terms in $F(x, \dot{x}, t)$ are included in $\mathbf{f}(t)$ as

$$\mathbf{f}(t) = \begin{bmatrix} 0 \\ -F(x, \dot{x}, t) \end{bmatrix} = \begin{bmatrix} 0 \\ -\beta_3 x^3 - \Gamma\dot{x} + F_0 \cos(\omega t + \theta)/m \end{bmatrix}. \tag{2.4}$$

However, when attempting to solve eqn (1.28) with this $\mathbf{f}(t)$, we note that $\mathbf{x}(t)$ is now subject to forcing by x and $\dot{x}$, yielding a problem that can only be solved self-consistently. Specifically, we do not know what frequency components are included in $x(t)$. Any combination of components is possible, as we emphasize by writing the Fourier transform

$$x(t) = \int_0^\infty x(\omega')e^{-i\omega' t} d\omega'\,, \tag{2.5}$$

with the Fourier amplitudes $x(\omega')$ and the integration variable ω'. Accordingly, the time integral in eqn (1.28) now contains all possible combinations within

$$x(t)^3 = \int_0^\infty \int_0^\infty \int_0^\infty x(\omega')x(\omega'')x(\omega''')e^{-i\omega' t}e^{-i\omega'' t}e^{-i\omega''' t} d\omega' d\omega'' d\omega'''\,, \tag{2.6}$$

which mix with the basis solutions in the transfer matrix Φ. This mixing between three components in $\mathbf{f}(t)$ and one in Φ is coined **four-wave mixing**. Nonlinearities lead to mixing between different oscillation frequencies because the products of exponential functions $\prod_j e^{-i\omega_j t}$ have components at all linear combinations of their frequencies ω_i. For a general polynomial potential of order n, we end up with a resonant effect at ω_0 if the product of $n-1$ cosines has a component at ω_0 (the resonant oscillation corresponds to the nth wave). In the following, we will learn how to treat an EOM with nonlinear terms.

2.1.2 The Poincaré–Lindstedt Method

One may immediately guess from Fig. 2.1 that $\beta_3 < 0$ reduces the frequency of unforced oscillations, while $\beta_3 > 0$ increases it. For a confirmation of this intuition, we employ the Poincaré–Lindstedt perturbative method [91, 92] to solve the homogeneous equation

$$\ddot{x} + \omega_0^2 x + \epsilon\beta_3 x^3 = 0\,, \tag{2.7}$$

where $0 < \epsilon \ll 1$ is a small parameter. Expecting that the nonlinearity leads to an amplitude-dependent frequency, we express frequency and time in stretched frames as $\omega_0\nu$ and $\tau = \nu t$, respectively, and expand both x and ν in orders of ϵ,

$$x = x_{(0)} + x_{(1)}\epsilon + x_{(2)}\epsilon^2 + \dots \tag{2.8}$$

$$\nu = 1 + k_{(1)}\epsilon + k_{(2)}\epsilon^2 + \dots\,. \tag{2.9}$$

Our equation now reads

$$\nu^2\frac{d^2x}{d\tau^2} + \omega_0^2 x + \epsilon\beta_3 x^3 = 0\,. \tag{2.10}$$

We can sort the ensuing terms according to their order in ϵ. The lowest-order solution is obtained by only accepting terms that do not contain ϵ at all,

$$\frac{d^2x_{(0)}}{d\tau^2} + \omega_0^2 x_{(0)} = 0\,. \tag{2.11}$$

This differential equation corresponds to the harmonic oscillator with the usual solution $x_{(0)} = X\cos(\omega_0\nu t + \phi)$ with $\nu = 1$ in the lowest-order approximation. In the absence of external forcing or particular initial conditions, we can choose $\phi = 0$. The first-order correction contains terms that are linear in ϵ,

$$\frac{d^2x_{(1)}}{d\tau^2} + 2k_{(1)}\frac{d^2x_{(0)}}{d\tau^2} + \omega_0^2 x_{(1)} + \beta_3 x_{(0)}^3 = 0\,. \tag{2.12}$$

Inserting the solution of the harmonic oscillator, $x_{(0)} = X\cos(\omega_0\tau)$, into eqn (2.12) and using the identity $\cos(\omega_0\tau)^3 = \frac{3}{4}\cos(\omega_0\tau) + \frac{1}{4}\cos(3\omega_0\tau)$, we get

$$\frac{d^2x_{(1)}}{d\tau^2} + \omega_0^2 x_{(1)} = \left[2\omega_0^2 X k_{(1)} - \frac{3X^3\beta_3}{4}\right]\cos(\omega_0\tau) - \frac{X^3\beta_3}{4}\cos(3\omega_0\tau)\,. \tag{2.13}$$

Equation (2.13) describes a resonantly driven, undamped oscillator. This system will diverge under the influence of the resonant force term unless we preserve the perturbative expansion by enforcing the so-called secular condition

$$2\omega_0^2 X k_{(1)} = \frac{3X^3\beta_3}{4}\,. \tag{2.14}$$

The secular condition fixes $k_{(1)}$ and thus provides us with the correction to the resonance frequency

$$\nu = 1 + \frac{3}{8}\frac{\beta_3}{\omega_0^2}X^2. \tag{2.15}$$

In agreement with our intuition, positive (negative) β_3 will lead to a higher (lower) resonance frequency.

A second important result is that the Duffing nonlinearity leads to **amplitude corrections** at ω_0 and gives rise to **harmonics** at $3\omega_0$. Based on eqn (2.12), we now want to estimate the amplitudes of these contributions. This first-order correction is an oscillator in itself that is driven by the force term

$$\frac{F}{m} = -\frac{X^3\beta_3}{4}\cos(3\omega_0\tau)\,. \tag{2.16}$$

From the premise that we are dealing with a perturbative expansion, we impose the boundary conditions

$$x_{(1)}(0) = \dot{x}_{(1)}(0) = 0\,. \tag{2.17}$$

We can now employ eqn (1.28), where the force at $3\omega_0$ in eqn (2.16) generates an oscillation at ω_0 via four-wave mixing, leading to

$$x_{(1)}(t) = \frac{X^3\beta_3}{32\omega_0^2\nu^2}\left[\cos\left(3\omega_0\tau\right) - \cos\left(\omega_0\tau\right)\right]\,. \tag{2.18}$$

The full solution to first order reads

$$x = X\cos\left(\omega_0\nu t\right) + \frac{X^3\beta_3}{32\omega_0^2}\left[\cos\left(3\omega_0\nu t\right) - \cos\left(\omega_0\nu t\right)\right]\,, \tag{2.19}$$

with the amplitude-dependent correction to the frequency given by eqn (2.15).

2.1.3 The Averaging Method

The Poincaré–Lindstedt method allows us to derive a relationship between the frequency of the free oscillator and its amplitude X. Commonly, however, we are interested in the EOM of the Duffing resonator subject to external driving as in eqn (2.1), and we want to track not only the long-time limit of the system, but also its transient dynamics. For this, we use the Krylov–Bogolyubov averaging method [91, 93, 94]. For readers interested in the somewhat similar **secular perturbation technique**, we recommend Ref. [45].

We return to the EOM in eqn (2.1),

$$\ddot{x} + \omega_0^2 x + F(x,\dot{x},t) = 0\,, \tag{2.20}$$

with $F(x,\dot{x},t)$ defined in eqn (2.3). Formally, the averaging procedure requires that all terms within $F(x,\dot{x},t)$ are of order ϵ (with $0 < \epsilon \ll 1$), and that we find corrections perturbatively in ϵ. We therefore rewrite eqn (2.3) as

$$F(x,\dot{x},t) = \epsilon\tilde{F}(x,\dot{x},t) = \epsilon\left(\tilde{\beta}_3 x^3 + \tilde{\Gamma}\dot{x} - \tilde{F}_0\cos(\omega t + \theta)/m\right)\,, \tag{2.21}$$

with the rescaled coefficients $\tilde{\beta}_3 = \beta_3/\epsilon$, $\tilde{\Gamma} = \Gamma/\epsilon$, and $\tilde{F}_0 = F_0/\epsilon$.

The idea behind the averaging method is that the weak influence of the perturbation $\epsilon\tilde{F}$ will not have an appreciable effect within one oscillation period. In other words, all relevant information for the long-time dynamics appears in the slowly changing amplitude and phase. Under these conditions, it is reasonable to assume that the solution will be close to that of the harmonic oscillator, $x(t) = x_{\text{ini}}e^{i\omega t}$, which would be exactly correct for $\epsilon = 0$ and $\omega = \omega_0$. Using the variational ansatz from eqn (1.23), we obtain the equation

$$\begin{bmatrix} x(t) \\ \dot{x}(t) \end{bmatrix} = \mathbf{x}(t) = W_P(t)\mathbf{d}(t) = \begin{bmatrix} \cos(\omega t) & -\sin(\omega t) \\ -\omega\sin(\omega t) & -\omega\cos(\omega t) \end{bmatrix} \begin{bmatrix} U(t) \\ V(t) \end{bmatrix}, \tag{2.22}$$

where the quadrature coefficients $U(t)$ and $V(t)$ are the entries of $\mathbf{d}(t)$, and $W_P(t)$ is a matrix that resembles the Wronskian but uses ω instead of ω_0; this transformation anticipates that the resonator mostly responds with the frequency of the drive.[1] The quadrature coefficients are explicitly time dependent to allow for slow transients of the amplitude and phase of the resonator. Equation (2.22) is also called the **van der Pol transformation.**

In eqns (1.23)–(1.28), we applied the variational ansatz to a linear differential equation and found an exact formula to predict $\mathbf{d}(t)$ and $\mathbf{x}(t)$ for arbitrary times. This is no longer generally possible when nonlinear terms are added to $\epsilon\tilde{F}(x, \dot{x}, t)$. However, we can still find an expression for $\dot{\mathbf{d}}(t)$ using an analogue to eqn (1.26). Here, we derive eqn (2.22) with respect to time and isolate $\dot{\mathbf{d}}$,

$$\dot{\mathbf{d}} = W_P^{-1}\dot{\mathbf{x}} - W_P^{-1}\dot{W}_P\mathbf{d}. \tag{2.23}$$

We now plug the expression for $\dot{\mathbf{x}}$ from eqn (1.18) and obtain

$$\begin{aligned} \begin{bmatrix} \dot{U} \\ \dot{V} \end{bmatrix} = \dot{\mathbf{d}} &= W_P^{-1} \begin{bmatrix} 0 \\ -\epsilon\tilde{F}(W_P\mathbf{d}, t) \end{bmatrix} + W_P^{-1} \begin{bmatrix} 0 & 1 \\ -\omega_0^2 & 0 \end{bmatrix} W_P\mathbf{d} - W_P^{-1}\dot{W}_P\mathbf{d} \\ &= W_P^{-1} \begin{bmatrix} 0 \\ -\epsilon\tilde{F}(W_P\mathbf{d}, t) \end{bmatrix} + W_P^{-1} \begin{bmatrix} 0 & 0 \\ \omega^2 - \omega_0^2 & 0 \end{bmatrix} W_P\mathbf{d} \\ &\equiv \epsilon\mathbf{F}_{\mathbf{d}}(\mathbf{d}, t), \end{aligned} \tag{2.24}$$

where we have dropped the explicit time dependencies for convenience and assume $|\omega^2 - \omega_0^2|$ to be of order ϵ. The notation $\epsilon\tilde{F}(W_P(t)\mathbf{d}(t), t)$ reflects the fact that all of the nonlinear terms in $\tilde{F}$ are now explicitly dependent on $W_P(t)\mathbf{d}(t)$ instead of $x(t)$ and $\dot{x}(t)$.

In a next step, we apply the **near-identity transformation** by writing $\mathbf{d}$ as a power series in ϵ,

$$\mathbf{d} = \mathbf{D}_0(t) + \epsilon\mathbf{D}_1(\mathbf{D}_0(t), t) + \epsilon^2\mathbf{D}_2(\mathbf{D}_0(t) + \epsilon\mathbf{D}_1(\mathbf{D}_0(t), t), t) + \ldots, \tag{2.25}$$

where $\mathbf{D}_0(t)$ is the lowest-order remaining time-dependence after the rotations, and where the *generating functions* $\mathbf{D}_i$ are yet unknown and depend iteratively on the predecessors in the hierarchy. We now proceed to insert eqn (2.25) into eqn (2.24),

[1] A more complete ansatz can be obtained through a method called *harmonic balance* [95, 96].

expand both sides in ϵ, and sort the result by orders of the expansion. Keeping only terms up to second order (and suppressing the functional dependencies), this procedure yields

$$(1 + \epsilon \mathbf{D}_1' + \epsilon^2 \mathbf{D}_2')\dot{\mathbf{D}}_0 = \epsilon \left[-\dot{\mathbf{D}}_1 + \mathbf{F_d} \right] + \epsilon^2 \left[-\dot{\mathbf{D}}_2 + \mathbf{F_d'} \mathbf{D}_1 \right] , \tag{2.26}$$

where we write $\mathbf{F_d'}$ to denote the matrix of partial derivatives with respect to $\mathbf{x}$. Multiplying by the inverse of $(1 + \epsilon \mathbf{D}_1' + \epsilon^2 \mathbf{D}_2')$ yields up to second order

$$\dot{\mathbf{D}}_0 = \epsilon \left[-\dot{\mathbf{D}}_1 + \mathbf{F_d} \right] + \epsilon^2 \left[-\dot{\mathbf{D}}_2 + \mathbf{F_d'} \mathbf{D}_1 - \mathbf{D}_1' \left(-\dot{\mathbf{D}}_1 + \mathbf{F_d} \right) \right] . \tag{2.27}$$

We now need to select appropriate $\mathbf{D}_i$. Our original motivation is to find an integral expression for $\mathbf{d}(t)$ similar to eqn (1.27), with $\mathbf{F_d}$ replacing $\mathbf{f}$. The problem we face in doing so is that $\mathbf{F_d}$ can include parts that do not average out over a period, leading to diverging integrals over arbitrary (infinite) times. To avoid this problem, we choose the $\mathbf{D}_i$ in such a way that they exactly cancel these diverging parts, that is [91, 93]

$$\begin{aligned} \dot{\mathbf{D}}_1 &= \mathbf{F_d} - \frac{1}{T_{\text{av}}} \int_0^{T_{\text{av}}} \mathbf{F_d} dt , \\ \dot{\mathbf{D}}_2 &= \mathbf{G_d} - \frac{1}{T_{\text{av}}} \int_0^{T_{\text{av}}} \mathbf{G_d} dt , \end{aligned} \tag{2.28}$$

where we defined $\mathbf{G_d} \equiv \mathbf{F_d'} \mathbf{D}_1 - \mathbf{D}_1' \left(-\dot{\mathbf{D}}_1 + \mathbf{F_d} \right)$. By this choice, we achieve that only the time-averaged terms remain in eqn (2.26), yielding the averaging equation

$$\begin{bmatrix} \dot{u} \\ \dot{v} \end{bmatrix} \equiv \dot{\mathbf{D}}_0 = \epsilon \frac{1}{T_{\text{av}}} \int_0^{T_{\text{av}}} \left[\mathbf{F_d}(\mathbf{D}_0(t), t) + \epsilon \mathbf{G_d} + \ldots \right] dt , \tag{2.29}$$

where we defined the time-averaged entries of D_0 as $u \equiv \bar{U}$ and $v \equiv \bar{V}$. Note that to lowest order, our result is equivalent to what we would have gained by applying the **averaging theorem**, which states that it is possible to replace the right-hand side of eqn (2.24) by its average over a time period T_{av} for well-behaved systems [93]. The value of this *averaging time* is typically chosen as the periodicity of $\mathbf{F_d}(\mathbf{D}_0, t)$, or (if no such term is present) as the natural period of the system. Note that faster dynamics of the system are captured not in $\mathbf{D}_0$ but in $\mathbf{D}_1$.

It may appear unnecessary to apply the near-identity transformation and search for suitable generating functions; after all, to lowest order the result is the same as what the averaging theorem yields in a simpler way. For first-order averaging, we could indeed jump from eqn (2.24) to eqn (2.29), neglect the ϵ^2 contribution with this direct argument, and save ourselves some work. However, we will see in Section 2.2 that some systems require second-order averaging where the near-identity transformation is essential.

2.1.4 First-Order Averaging of the Quartic Potential

For the driven Duffing resonator, cf. eqn (2.1), the first-order averaged solutions in eqn (2.29) are

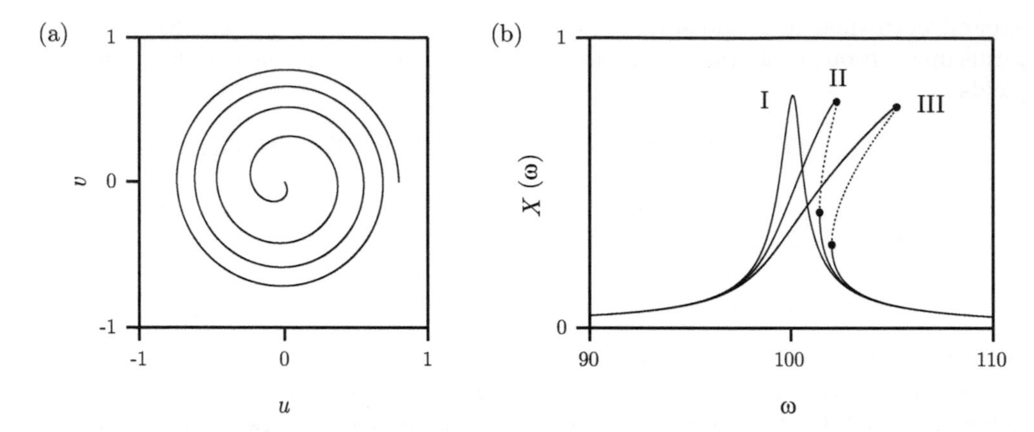

Fig. 2.2 (a) Ringdown of a Duffing resonator in the rotating frame calculated with the averaging method. The frame rotates at ω_0 with $F_0 = 0$, $m = 1$, $\omega_0 = 100$, $\Gamma = 1$, and $\beta_3 = 1$. Initial conditions are $x_i = 0.8$ and $\dot{x}_i = 0$. The trace was simulated over a time of $t = 1000$. (b) Steady-state response of driven Duffing resonator. Parameters are as in (a) except for $F_0 = 80$ and $\beta_3 = 1$. Examples I corresponds to $\beta_3 = 100$, II to $\beta_3 = 2000$, and III to $\beta_3 = 5000$. Unstable solution branches are shown as dotted lines and bifurcation points are indicated as large dots.

$$\dot{u} = -\frac{u\Gamma}{2} - \frac{3u^2 v\beta_3}{8\omega} - \frac{3v^3\beta_3}{8\omega} + \frac{v\left(\omega^2 - \omega_0^2\right)}{2\omega} + \frac{F_0 \sin(\theta)}{2m\omega}, \tag{2.30}$$

$$\dot{v} = -\frac{v\Gamma}{2} + \frac{3v^2 u\beta_3}{8\omega} + \frac{3u^3\beta_3}{8\omega} - \frac{u\left(\omega^2 - \omega_0^2\right)}{2\omega} - \frac{F_0 \cos(\theta)}{2m\omega}. \tag{2.31}$$

The slow-flow eqns (2.30) and (2.31) can be solved numerically to observe the time evolution of u and v from any starting condition. In addition, these equations are conceptually important because they transport us into the **rotating frame** where oscillations are perceived only relative to a specified rotation of the observer, similar to what occurs in the lab when using a lock-in amplifier. A system oscillating at ω_0 with stable amplitude and phase becomes a single point in the u/v phase space when it is observed (or *demodulated*) at this frequency. Slow changes in amplitude and phase, therefore, correspond to a slow evolution in the (rotated and averaged) phase space.

In Fig. 2.2(a), we plot the ringdown curve of a Duffing resonator in a frame rotating at ω_0. Interestingly, the phase revolves slowly while the amplitude decreases. This is due to the amplitude dependence of the resonance frequency of the system that we also calculated with the Poincaré–Lindstedt method, cf. eqn (2.15). Since $\beta_3 > 0$, the oscillation frequency is initially above ω_0. The variations in phase that we observe are nothing but deviations of the oscillation frequency from the rotating frame. It is apparent that the slow oscillations decelerate for smaller amplitudes where the frequency converges toward ω_0.

Moving beyond the non-driven case, we can now study the stationary response of the Duffing oscillator to a harmonic external force. To this end, we impose $\dot{u} = \dot{v} = 0$

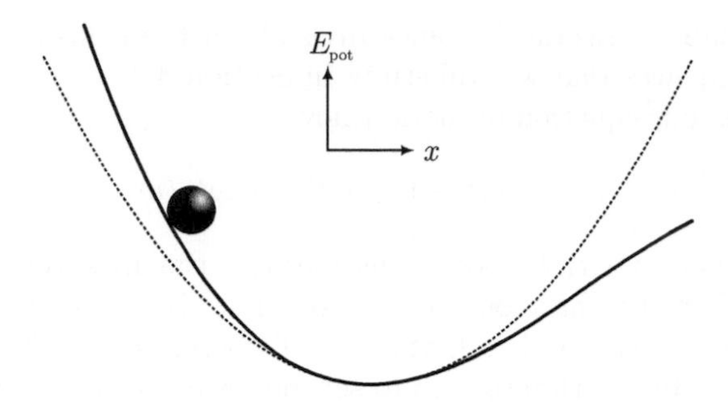

Fig. 2.3 A quadratic nonlinearity in the spring constant corresponds to a cubic potential term. A harmonic potential is shown as a dotted line for reference.

and need to solve the resulting coupled third-order polynomial equations. Technically, we solve the equations for complex u and v, but keep only physical (real) solutions. We furthermore sort the solutions according to their stability, cf. Section 1.3.2. In Section 3.1.3, we provide a rigorous demonstration of this procedure. Points where the number of physical solutions change are called **bifurcation points**.

We plot the amplitude response of the resonator for various nonlinearities in Fig. 2.2(b). Below a critical nonlinearity β_3, the response is always single-valued, that is, there exists exactly one stable solution. For sufficiently large β_3, however, there is a frequency range where two stable solutions exist together with a third unstable one. In an experiment, the various stable solutions can be probed by performing a frequency sweep by slowly (adiabatically) changing the driving frequency ω. Depending on the sweep direction (upward or downward in frequency), different solution branches are sampled. This phenomenon is known as **hysteresis**.

2.2 The Cubic Potential

After discussing the consequences of adding a quartic term to the resonator potential, we will now consider a cubic term (and thus a quadratic restoring force). We will see that the effect is very similar to that of the Duffing nonlinearity. In the slow-flow picture, we will even express the quadratic force term as an effective Duffing nonlinearity.

What do we expect to find when adding any antisymmetric contribution to the resonator potential? Clearly, the potential will become steeper on one side and shallower on the other. As for the quartic potential, we choose to ignore the regime of oscillations that are large enough to lead to instability. For relatively small oscillation amplitudes, the asymmetry leads to an increase in the time spent on one half of the potential (the right half in Fig. 2.3), while the time spent on the other half is decreased. Interestingly, the sign of the cubic potential term does not matter for the periodicity of the full cycle. We can already guess that the resonance frequency decreases with amplitude because the effect of the shallow side is dominant.

Within each cycle, the nonlinearity leads to deformations from the harmonic motion that give rise to a Fourier component at **twice** the fundamental oscillation, in analogy to what we have obtained for a quartic potential in eqn (2.19). This **three-wave**

mixing within the cubic potential becomes invisible in the slow-flow equations but has important consequences that we will study in Section 3.2.

In the driven case, our equation of motion now reads

$$\ddot{x} + \omega_0^2 x + \beta_2 x^2 + \Gamma \dot{x} = F_0 \cos(\omega t)/m. \tag{2.32}$$

When solving this equation with the averaging method, the first-order correction produces no deviation from the harmonic resonator. This is due to the fact that the half-cycle period corrections cancel to first order. To perceive the effect of β_2, we apply the averaging procedure with terms up to second order in ϵ in eqn (2.26) and with both generating functions from eqn (2.28).

2.2.1 Second-Order Averaging of the Cubic Potential

The second-order slow-flow equations for eqn (2.32) take the form

$$\dot{u} = -\frac{u\Gamma}{2} + \frac{5u^2 v \beta_2^2}{12\omega^3} + \frac{5v^3\beta_2^2}{12\omega^3} + \frac{v\left(\omega^2 - \omega_0^2\right)}{2\omega} + \frac{F_0 \sin(\theta)}{2m\omega}, \tag{2.33}$$

$$\dot{v} = -\frac{v\Gamma}{2} - \frac{5v^2 u \beta_2^2}{12\omega^3} - \frac{5u^3\beta_2^2}{12\omega^3} - \frac{u\left(\omega^2 - \omega_0^2\right)}{2\omega} - \frac{F_0 \cos(\theta)}{2m\omega}. \tag{2.34}$$

When comparing these equations with eqns (2.30) and (2.31), we notice an interesting connection. Indeed, the two sets of equations are identical when replacing the Duffing nonlinearity β_3 in eqns (2.33) and (2.34) with $-\frac{10}{9}\beta_2^2\omega^{-2}$. In general, we can show that a system with both types of nonlinearities can in the rotating frame be described through an effective frequency-dependent Duffing nonlinearity that we can express as

$$\beta = \beta_3 - \frac{10}{9}\frac{\beta_2^2}{\omega^2}. \tag{2.35}$$

We see from eqn (2.35) that quadratic nonlinear forces will always lead to a negative effective Duffing term β, and thus to a decrease of the resonance frequency with increasing amplitude. In the following, we will generally use a term βx^3 in the force equation to account for both nonlinearities in the rotating frame. We should remember, however, that this nonlinear coefficient is itself frequency dependent if it contains a contribution from β_2. This frequency dependence can lead to interesting phenomena beyond the physics discussed in this text.

Chapter summary

- In Chapter 2, we extend our classical, deterministic resonator model to include cubic and quartic potential terms, that is, quadratic and cubic restoring forces with coefficients β_2 and β_3, respectively.
- We write the equation of motion of the driven and damped oscillator with a cubic restoring force, which is the so-called **Duffing nonlinearity**, cf. eqn (2.1). With the **Poincaré–Lindstedt method**, we find that the Duffing nonlinearity causes an amplitude-dependent shift of the resonance frequency, as well as harmonics at integer multiples of the resonance frequency, cf. eqn (2.19).
- To investigate the response of the Duffing resonator to a near-resonant external force, we employ the **averaging method**. In this method, features that change fast in time are averaged out, and we only retain the slowly varying amplitude or phase, or, equivalently, the quadratures u and v in a phase space rotating at a frequency $\omega \approx \omega_0$, cf. eqn (2.29). We refer to these coupled first-order equations as the slow-flow equations.
- The Duffing resonator can be treated with the first-order averaging method, cf. eqns (2.30) and (2.31). These equations allow us to calculate the transient or long-time limit responses of the system, cf. Fig. 2.2.
- When including a quadratic restoring force, the first-order averaging method is insufficient. We therefore proceed to second-order averaging, cf. eqns (2.33) and (2.34). From these results, we conclude that the effect of the quadratic nonlinearity can be reduced (in the rotating frame) to a rescaled Duffing coefficient β, cf. eqn (2.35).

Exercises

Check questions:

(a) Draw the potential well of a Duffing oscillator, for both $\beta_3 > 0$ and $\beta_3 < 0$. How does the Duffing term change the relationship between displacement and potential energy? How does this change the resonance frequency as a function of displacement and for different signs of β_3?

(b) Draw the potential well of an oscillator with a quadratic nonlinearity, for both $\beta_2 > 0$ and $\beta_2 < 0$. How does this nonlinear term change the relationship between displacement and potential energy? How does this change the resonance frequency as a function of displacement and for different signs of β_2?

(c) Can you explain in a simple way why nonlinear terms generate harmonics? What gives rise to the first-order amplitude correction at ω_0 in the Poincaré–Lindstedt method, cf. eqn (2.18)?

(d) What conditions lead to the observation of a hysteresis in a frequency sweep? What is the role of bifurcation points for hysteresis? Can you guess from simple considerations how the stability of a system evolves close to a bifurcation point?

Tasks:

2.1 Familiarize yourself with the added features in the Jupyter Notebook **Python Example 2**, especially the rotating-frame representation. This function allows you to follow the slow quadratures **u_av** and **v_av** in phase space (in a computationally inefficient way).

2.2 Set $\lambda = 0$ (“**lam=0**”), $F_1 = 0$ (“**F1=0**”), and $\eta = 0$ (“**eta=0**”), and explore the effect of the nonlinear terms **beta3** and **beta2** individually. Use $\omega_0 = 1$, $m = 1$, and $\Gamma = 0.01$ as starting values. Can you detect the influence of the nonlinearities in a ring-down experiment (**F0=0**)? Can you generate a situation where harmonics become visible in the FFT? What is the new effect you observe in phase space due to the nonlinearities?

2.3 Perform a frequency sweep in the presence of a Duffing nonlinearity **beta3** (with **beta2=0**). Use $\omega_0 = 1$ and $\Gamma = 0.01$ as starting values. A large sampling time step makes the simulation more efficient — roughly 10 points per oscillation ($2\pi/\omega_0$) are enough for weak nonlinearities. Increase this value if you observe numerical artifacts.

2.4 Test different driving forces (**F0**) in both sweep directions and extract for each sweep the frequency where the amplitude response reaches a maximum. These points form the so-called *backbone function* of the Duffing resonator. They are a physical manifestation of the amplitude-dependency of the resonance frequency that we found in the Poincaré–Lindstedt method. Can you relate your sweep results to eqn (2.15) to extract your Duffing term from the simulation results?

2.5 Perform a frequency sweep with a negative Duffing term (**beta3<0**) and a sufficiently large force to observe a hysteresis. Using eqn (2.35), calculate what value of **beta2** is needed to achieve the same effect. Using this value of **beta2** in combination with a positive Duffing nonlinearity **beta3**, what do you observe? Can you explain the observation with the potential well picture?

2.6 Familiarize yourself with the Jupyter Notebook **Python Example 3**. Here, we directly operate with slow-flow equations, which is much more efficient than the nonrotating frame. The frequency in this tool sweeps continuously from **omega1** to **omega2** (set **omega2=omega1** for a simple run). Set **sigma_D=0** for now (this will later quantify our force noise). This notebook requires the "**sdeint**" package.

2.7 Repeat some of the numerical experiments from above with **Python Example 3**. This code is much more efficient to observe the long-time limit of the amplitude and phase response. As a calibration exercise, test that you obtain the same result as with **Python Example 2** in a frequency sweep with $\omega_0 = 1$, $\Gamma = 0.01$, and $F_0 = 1$ without any nonlinearity.

3

Degenerate Parametric Pumping

In Chapter 1, we briefly discussed the influence of parametric pumping, namely periodic modulations of the spring constant at a frequency ω_p. For the important case of $\omega_p \approx 2\omega_0$, such a modulation leads to so-called degenerate parametric pumping around the lowest Arnold tongue, see Fig. 1.6. If the modulation is sufficiently strong, we found that it can destabilize the harmonic resonator, but we were not yet able to predict what happens beyond that point. In Chapter 2, we familiarized ourselves with the Duffing nonlinearity and with the averaging method. Equipped with this knowledge, we now turn again to the parametric resonator and seek a general description of its behavior below and above the modulation threshold.

3.1 The Nonlinear Parametric Resonator

The equation of motion we consider here is

$$\ddot{x} + \omega_0^2 \left[1 - \lambda \cos\left(2\omega t + \psi\right)\right] x + \beta x^3 + \Gamma \dot{x} + \eta x^2 \dot{x} = \frac{F_0}{m} \cos\left(\omega t + \theta\right) , \tag{3.1}$$

where λ is again the modulation depth as in the Hill or Mathieu equations, cf. eqns (1.32) and (1.31), and where we have added a nonlinear damping term with coefficient η. As we will see, nonlinear damping is important in order to understand one particular feature of the parametric resonator. For a better notion of *detuning* of the parametric drive, we use the notation $2\omega \equiv \omega_p$. Equation (3.1) can be treated with the averaging method using an averaging time of $T_{\text{av}} = 2\pi/\omega$ to obtain

$$\begin{aligned}
\dot{u} = &-\frac{u\Gamma}{2} - \frac{\left(u^3 + uv^2\right)\eta}{8} - \frac{3\left(u^2 v + v^3\right)\beta}{8\omega} + \frac{v\left(\omega^2 - \omega_0^2\right)}{2\omega} \\
&+ \frac{\lambda\omega_0^2}{4\omega}\left(u \sin\psi - v\cos\psi\right) + \frac{F_0 \sin\theta}{2m\omega} ,
\end{aligned} \tag{3.2}$$

$$\begin{aligned}
\dot{v} = &-\frac{v\Gamma}{2} - \frac{\left(v^3 + u^2 v\right)\eta}{8} + \frac{3\left(uv^2 + u^3\right)\beta}{8\omega} - \frac{u\left(\omega^2 - \omega_0^2\right)}{2\omega} \\
&- \frac{\lambda\omega_0^2}{4\omega}\left(u\cos\psi + v\sin\psi\right) - \frac{F_0\cos\theta}{2m\omega} .
\end{aligned} \tag{3.3}$$

By construction, this result allows us to study the resonator response at ω to a parametric pump acting at 2ω, which we refer to as **periodicity doubling**. Responses at higher frequencies, which do exist for strong pumping, are averaged out. For the steady-state condition $\dot{u} = \dot{v} = 0$, eqns (3.2) and (3.3) can be cast into a single equation for the real amplitude $X^2 = u^2 + v^2$, namely [94]

$$\left[-\frac{\lambda^2}{4} + A^2 + B^2\right]^2 X^2 = \frac{F_0^2}{m^2\omega_0^4}\left[\frac{\lambda^2}{4} + A^2 + B^2 + \lambda A \sin(2\theta - \psi) + \lambda B \cos(2\theta - \psi)\right], \quad (3.4)$$

where we define

$$A = \frac{\Gamma\omega}{\omega_0^2} + \frac{\eta\omega X^2}{4\omega_0^2},$$
$$B = 1 - \frac{\omega^2}{\omega_0^2} + \frac{3\beta X^2}{4\omega_0^2}. \quad (3.5)$$

Equation (3.4) has been left very general in order to be applicable to a wide range of situations. For instance, we can use it to obtain a solution for the amplitude of the driven Duffing resonator in the presence of nonlinear damping. For $\lambda = 0$, eqn (3.4) simplifies to

$$X^2 = \frac{F_0^2}{m^2\omega_0^4}\frac{1}{A^2 + B^2}. \quad (3.6)$$

For $F_0 = 0$ but with $\lambda > 0$, the equation allows us to determine the regions of stability and instability of the parametric resonator, and the hysteretic amplitude response in the latter case. It even enables us to understand parametric amplification and symmetry breaking when a parametric pump and an external force are present simultaneously, as we discuss in the following.

3.1.1 Effective Damping Below Threshold

We begin with the case of a parametric linear resonator, which arises for $\beta = \eta = F_0 = 0$ and for small enough λ. Since we have not yet done so, let us first define what we mean by "small λ." For the conditions stated here, we can simplify eqn (3.4) to

$$X^2\left[(\Gamma\omega)^2 - \left(\frac{\lambda\omega_0^2}{2}\right)^2 + \left(\omega_0^2 - \omega^2\right)^2\right]^2 = 0. \quad (3.7)$$

When the parametric pump is applied resonantly, $\omega = \omega_0$, what is left in the square root is a reduced damping term of the form

$$\Gamma^2 - \frac{\lambda^2\omega_0^2}{4}. \quad (3.8)$$

The reduced damping term becomes zero when $\lambda = \lambda_{\text{th}} = 2/Q$. This is the **parametric instability threshold** on resonance. Even though $X = 0$ is always a solution to eqn (3.4), beyond the threshold it is no longer a stable solution in the sense that

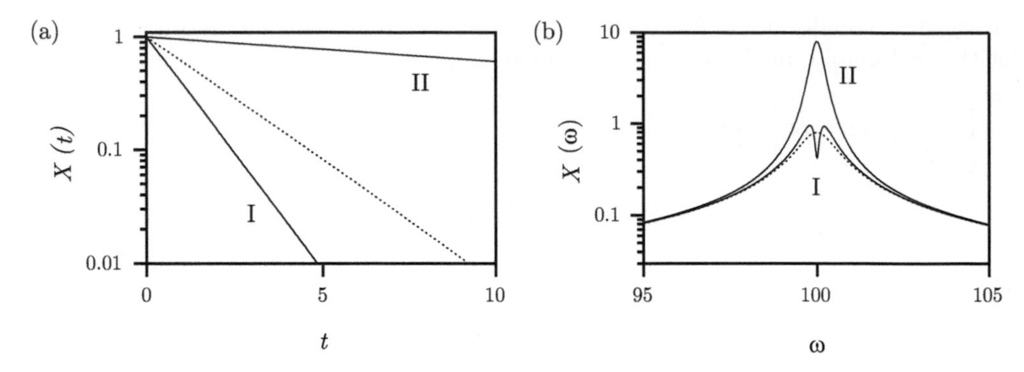

Fig. 3.1 (a) Ringdown of a harmonic oscillator in the presence of parametric pumping below threshold and on resonance. Trace I corresponds to $\psi = -\pi/2$ and II to $\psi = \pi/2$. Both curves were simulated from the solutions of eqns (3.2) and (3.3) for $F_0 = \beta = \eta = 0$, $\omega_0 = 100$, $\Gamma = 1$, $m = 1$, and $\lambda = 0.9\lambda_{\rm th}$. A ringdown with $\lambda = 0$ is shown as a dotted line for comparison. (b) Steady-state response of a driven, harmonic resonator with parametric pumping. All parameters are chosen to be the same as in (a) except for $F_0 = 80$ and $\theta = 0$. The response is shown on a logarithmic scale to make small features more visible.

any small amplitude will be exponentially amplified over time. In this section, we will consider only $\lambda < \lambda_{\rm th}$.

As long as $X = 0$, a reduction in damping has no consequences. When studying the effect of parametric pumping below threshold, we therefore require nonzero starting conditions or $F_0 \neq 0$, and we must consider the role of the parametric phase ψ in the full eqn (3.4). In Fig. 3.1(a), we show ringdown curves calculated from the solutions of eqns (3.2) and (3.3) for two different values of ψ. Depending on the phase of the pump relative to that of the undriven oscillation, the ringdown can be significantly faster or slower compared to the case without any parametric pump. This can be expressed in terms of an effective damping rate $\Gamma_{\rm eff}$ or effective quality factor $Q_{\rm eff}$, whose maximum and minimum values we can calculate from eqn (3.4) for $\omega \approx \omega_0$ and $\beta = \eta = 0$. We find

$$\Gamma_{+/-} = \Gamma \pm \frac{\lambda\omega_0^2}{2\omega} \approx \Gamma \pm \frac{\lambda\omega_0}{2}, \tag{3.9}$$

$$Q_{+/-} = \frac{\omega_0}{\Gamma_{+/-}} \approx \left(\frac{1}{Q} \mp \frac{\lambda}{2}\right)^{-1}. \tag{3.10}$$

The effective damping can be used to increase the steady-state response of the resonator to an external driving force. The gain of this **parametric amplification** is determined by λ and the phase difference $2\theta - \psi$, but it also depends sensitively on the frequency detuning relative to ω_0. Since the phase of the harmonic resonator undergoes a change of π around its resonance, the parametric phase ψ can only be chosen for optimal gain at one specific frequency ω. The curves in Fig. 3.1(b) therefore deviate from the simple shape of a resonance peak. When approaching the parametric instability threshold, the gain of the harmonic resonator becomes infinite and nonlinearities must be taken into account.

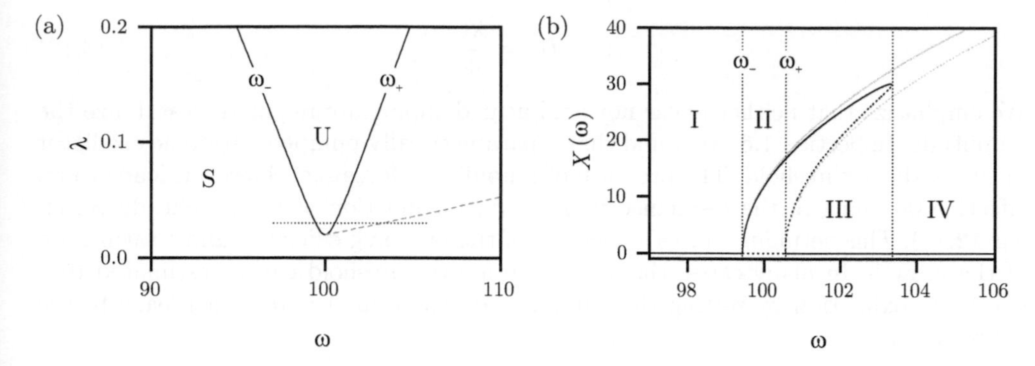

Fig. 3.2 (a) The boundary between stable (S) and unstable (U) behavior of the harmonic resonator with parametric pumping as calculated using eqn (3.11) for $\omega_0 = 100$, $m = 1$, and $\Gamma = 1$. (b) Steady-state responses of a resonator with and without nonlinear damping along the dotted horizontal line in (a). Solutions were calculated for the same parameters as in (a) and with $\lambda = 1.5\lambda_{\mathrm{th}}$, $\beta = 1$, and $F_0 = 0$. Stable and unstable solutions are shown as solid and dotted lines, respectively. Including $\eta = 0.002$ allows the stable and unstable solutions to merge in bifurcation points at $\omega \approx 103.5$ (black). Vertical dotted lines separate regions with constant number of stable solutions. The solutions for $\eta = 0$ merge at infinity (gray). A gray dashed line in (a) indicates the boundary between regions III and IV.

3.1.2 Parametric Pumping Above Threshold

When the parametric pumping becomes sufficiently strong, we enter the regime where the response of the harmonic resonator is no more stable. The region in a space spanned by ω and λ where this is true is often referred to as the **Arnold tongue** or instability lobe. The latter name can be confusing, because no instability actually takes place for nonlinear systems, see Fig. 1.6. Inside this region, we need to consider nonlinear effects, most prominently those caused by the Duffing potential, in order to understand the behavior of the parametric resonator.

Let us begin by determining the boundaries of the Arnold tongue from eqn (3.7). For small detuning of the pump frequency, the instability appears at higher pumping than λ_{th} and we find the simple form

$$\omega_\pm^2 = \omega_0^2\left(1 \pm \sqrt{\frac{\lambda^2}{4} - \frac{1}{Q^2}}\right), \tag{3.11}$$

where $\omega_\pm$ are the boundaries of the instability region for a given λ and quality factor $Q = \omega_0/\Gamma$, see Fig. 3.2(a). It is worth noting that to lowest order (mean-field treatment), β and η have no influence on these boundaries because the transition from stability to instability takes place at small oscillation amplitudes.

The long-time-limit solutions of a parametrically pumped resonator, as obtained by solving the steady-state limit of eqns (3.2) and (3.3), are shown in Fig. 3.2(b). In region I, the only solution in the absence of external driving is $X = 0$. In region II, this solution becomes unstable and there are two stable branches with equally large amplitudes given by

$$A^2 + B^2 = \frac{\lambda^2}{4}\,. \tag{3.12}$$

We emphasize that neither linear nor nonlinear damping are required to stabilize the amplitude. In Section 1.5, we found that a parametrically pumped harmonic oscillator diverges above threshold. The presence of a nonlinear force βx^3, however, leads to an effective detuning of the resonance frequency as a function of the amplitude X, cf. eqn (2.15). This detuning causes a decrease of the pumping efficiency and a saturation of the growth. In many cases, the amplitude above threshold can be estimated to a good approximation by setting the damping contribution $A = 0$, which leads to the simple form

$$X = \sqrt{\frac{2\lambda\omega_0^2}{3\beta}}\,. \tag{3.13}$$

Region II is defined by the boundaries of the Arnold tongue, cf. eqn (3.11). Beyond ω_+, the large-amplitude branches coexist with a third stable solution at $X = 0$, giving rise to a hysteresis in frequency sweeps: when sweeping from low to high frequencies, the system will remain in the phase state selected in region II throughout region III. When sweeping in the opposite direction, the system will remain in the zero-amplitude solution until the boundary of the Arnold tongue is reached at ω_+. For finite nonlinear damping, an increasing amplitude will lead to an increase in the overall damping, eventually causing Γ_{eff} to drop back to zero. This happens at bifurcation points where the large-amplitude branches merge with unstable branches and region III terminates. The nonlinear damping coefficient η is therefore directly responsible for the position of the boundary separating regions III and IV. For $\eta = 0$, region III terminates at infinite frequency and region IV is never reached. It is for this important effect that we included η in our EOM. In region IV, only $X = 0$ is stable. Regions I, III and IV make up the zone marked as S in Fig. 3.2(a); region II corresponds to U.

What about the phase of the parametric resonator? As long as $X = 0$, the resonator phase is naturally not defined. However, for the large-amplitude response, we find two possible oscillation phases. This is due to the period-doubling of the response relative to the parametric pump. There are two oscillations with opposite phases that are both solutions to eqn (3.12). In the absence of noise, which of the two **phase states** is realized depends solely on the initial conditions, see Fig. 3.3. In Section 5.3, we will understand how the presence of force noise can affect this behavior.

3.1.3 Stability and the Quasi-Potential Picture

So far, we have used the terms *stable* and *unstable* for the nonlinear parametric oscillator without clarifying what we mean thereby. In Section 1.3.2, we have briefly introduced the notion of an equilibrium point characterized by $\dot{\mathbf{x}} = 0$. There, we have defined a stable equilibrium point to be one that remains stable under small perturbations $\delta\mathbf{x}$. We can perform the same exercise for our averaged coordinates u and v in the rotating frame to establish how the system reacts to small perturbations $\delta u = u - u_{\text{eq}}$ and $\delta v = v - v_{\text{eq}}$. Here, u_{eq} and v_{eq} are the phase space coordinates of one particular equilibrium point where $\dot{u}_{\text{eq}} = \dot{v}_{\text{eq}} = 0$. For a stable equilibrium point, we demand that

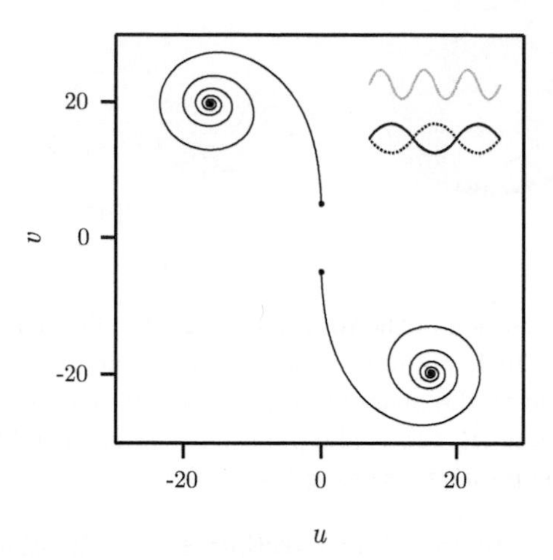

Fig. 3.3 Two ringups calculated from eqns (3.2) and (3.3). The curves start at the two black dots $(0/\pm 5)$ and evolve toward the two stable phase states. $\omega_0 = 100$, $m = 1$, $\Gamma = 1$, $\lambda = 5\lambda_{\mathrm{th}}$, $\psi = 0$, $\beta = 1$, and $F_0 = \eta = 0$. The frame rotates at $\omega = \omega_0$. The spiral motion is a consequence of the nonlinear coupling between u and v, see eqns (3.2) and (3.3). Insets show schematically the parametric pump (gray) and the two responses with opposite phase (solid and dotted black lines) as a function of time.

all characteristic exponents of the matrix equation governing δu and δv are negative. The linear evolution of the perturbations will follow an equation of the form

$$\begin{bmatrix} \delta\dot{u} \\ \delta\dot{v} \end{bmatrix} = M \begin{bmatrix} \delta u \\ \delta v \end{bmatrix} . \tag{3.14}$$

To find the relevant matrix M, we insert

$$\begin{bmatrix} u \\ v \end{bmatrix} = \begin{bmatrix} u_{\mathrm{eq}} \\ v_{\mathrm{eq}} \end{bmatrix} + \begin{bmatrix} \delta u \\ \delta v \end{bmatrix} \tag{3.15}$$

into eqns (3.2) and (3.3) and keep only terms that are linear in δu and δv, yielding equations

$$\begin{bmatrix} \delta\dot{u} \\ \delta\dot{v} \end{bmatrix} = \begin{bmatrix} f_u(u_{\mathrm{eq}} + \delta u, v_{\mathrm{eq}} + \delta v) \\ f_v(u_{\mathrm{eq}} + \delta u, v_{\mathrm{eq}} + \delta v) \end{bmatrix} \approx M(u_{\mathrm{eq}}, v_{\mathrm{eq}}) \begin{bmatrix} \delta u \\ \delta v \end{bmatrix} . \tag{3.16}$$

The resulting matrix M is the Jacobian matrix that contains the partial derivatives of $f_{1,2}$ with respect to δu and δv,

$$M = \begin{bmatrix} M_{11} & M_{12} \\ M_{21} & M_{22} \end{bmatrix} \tag{3.17}$$

with the entries corresponding to eqns (3.2) and (3.3), being

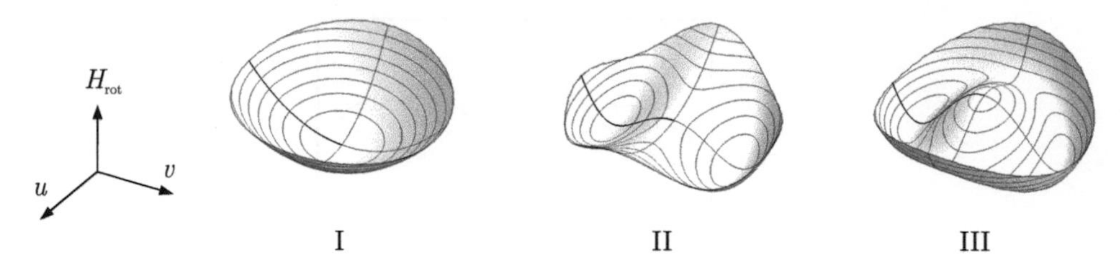

Fig. 3.4 Schematic illustration of the rotating-frame Hamiltonian H_{rot} from eqn (3.22) for the regions I, II, and III. In I, we find a single attractor at the origin; in II, there are two attractors that correspond to the phase states. The attractors are separated by an unstable saddle-point at the origin; in III, the saddle-point has become a stable attractor that appears as a maximum due to the positive detuning.

$$
\begin{aligned}
M_{11} &= \frac{\partial f_1}{\partial \delta u} = -\frac{6uv\beta + 4\Gamma\omega + \eta\omega(3u^2+v^2) - 2\lambda\omega_0^2 \sin\psi}{8\omega}, \\
M_{12} &= \frac{\partial f_1}{\partial \delta v} = -\frac{3\beta(u^2+3v^2) + 2uv\eta\omega + 4(\omega_0^2-\omega^2) + 2\lambda\omega_0^2 \cos\psi}{8\omega}, \\
M_{21} &= \frac{\partial f_2}{\partial \delta u} = \frac{3\beta(3u^2+v^2) - 2uv\eta\omega + 4(\omega_0^2-\omega^2) - 2\lambda\omega_0^2 \cos\psi}{8\omega}, \\
M_{22} &= \frac{\partial f_2}{\partial \delta v} = -\frac{-6uv\beta + 4\Gamma\omega + \eta\omega(u^2+3v^2) + 2\lambda\omega_0^2 \sin\psi}{8\omega}.
\end{aligned} \tag{3.18}
$$

In general, we assume that M has (normalized) eigenvectors a_j with corresponding eigenvalues μ_j, such that a system initialized at coordinates $u_{\text{ini}} = u_{\text{eq}} + \delta u_{\text{ini}}$ and $v_{\text{ini}} = v_{\text{eq}} + \delta v_{\text{ini}} ini$ evolves as

$$
\begin{bmatrix} u(t) \\ v(t) \end{bmatrix} = \begin{bmatrix} u_{\text{eq}} \\ v_{\text{eq}} \end{bmatrix} + \sum_j A_j a_j e^{\mu_j t} \tag{3.19}
$$

with coefficients A_j such that

$$
\sum_j A_j a_j = \begin{bmatrix} \delta u_{\text{ini}} \\ \delta v_{\text{ini}} \end{bmatrix}. \tag{3.20}
$$

Here we are only interested in the exponents μ_j which solve the characteristic equation

$$
\left| M - \begin{bmatrix} \mu & 0 \\ 0 & \mu \end{bmatrix} \right| = \mu^2 - \text{tr}(M)\mu + |M| = 0. \tag{3.21}
$$

A stable equilibrium point has only characteristic exponents with negative real parts, while positive real values indicate an unstable behavior. The general procedure is to first find all points that fullfil $\dot{u}_{\text{eq}} = \dot{v}_{\text{eq}} = 0$ at a given frequency, and then to determine the real parts of the characteristic exponents of the fluctuations for every one of these points. Repeating this procedure at all frequencies confirms the existence of 1, 2, 3, and 1 stable equilibrium points in the regions I–IV in Fig. 3.2, respectively.

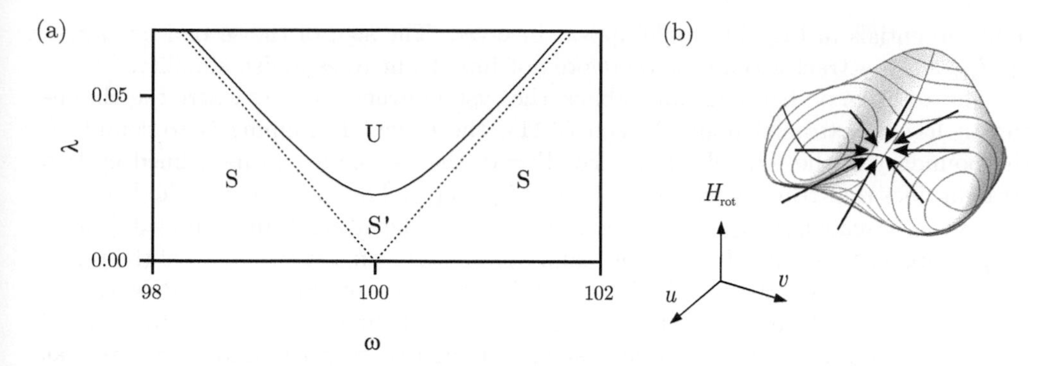

Fig. 3.5 (a) For $\Gamma = 0$, the tip of the Arnold tongue is sharp and extends down to $\lambda = 0$, as indicated by the dotted line. The region marked S' between the two lines is stabilized due to finite Γ that provides a centripetal force in eqns (3.2) and (3.3). The labels S and U refer to stable and unstable conditions for $u = v = 0$. (b) in the region S', the quasi-potential H_{rot} features a saddle point at the origin. Black arrows indicate the direction of the damping force.

The stability analysis in Section 1.3.2 was inspired by the dynamics of a damped particle in a potential well. It is therefore natural to ask if the stability conditions of the nonlinear parametric oscillator might as well be understood from some quasi-potential picture. Indeed, we can construct such a quasi-potential spanned by u and v from eqns (3.2) and (3.3) by setting $\Gamma = \eta = 0$ and defining a rotating-frame Hamiltonian

$$\begin{aligned} H_{\text{rot}} =& -\frac{3\beta}{32\omega}\left(2u^2v^2 + u^4 + v^4\right) + \frac{\omega^2 - \omega_0^2}{4\omega}\left(u^2 + v^2\right) \\ &+ \frac{\lambda\omega_0^2}{4\omega}\left(uv\sin\psi + \frac{u^2 - v^2}{2}\cos\psi\right) + \frac{F_0}{2m\omega}\left(u\cos\theta + v\sin\theta\right) \end{aligned} \tag{3.22}$$

that allows us to obtain eqns (3.2) and (3.3) as

$$\dot{u} = \frac{\partial H_{\text{rot}}}{\partial v} \qquad \dot{v} = -\frac{\partial H_{\text{rot}}}{\partial u}. \tag{3.23}$$

In Fig. 3.4 we plot H_{rot} for the regions I, II, and III in Fig. 3.2 and for $F_0 = 0$. In region I, the resonator is driven below the parametric threshold and exhibits a single stable solution at the origin. At the threshold $\lambda = \lambda_{\text{th}}$, the number of stable solution changes from 1 to 2, in analogy with a second-order phase transition at zero temperature [97]. Within region II, the two phase states are approximately located at the minima of two wide attractor pools with a saddle-point between them. The transition between regions II and III is marked by the transformation of this saddle-point into a maximum.

In contradiction to our intuition from nonrotating (equilibrium) potentials, the maximum in III is itself a stable attractor. To understand this, remember that potential energy is relative in the rotating frame. Depending on $\omega^2 - \omega_0^2$, the curvature of H_{rot} can be positive, negative, or zero. For instance, if we inverted the sign of β, region I and III would correspond to positive and negative detuning, respectively, and the quasi-potentials in Fig. 3.4 would appear inverted. The sign of the second derivative of H_{rot} at an attractor center is therefore not important to establish stability.

The addition of damping can stabilize the system even at points where the Hamiltonian has no extremal point. In eqn (3.11), the effect of damping is to round off the boundary of the Arnold tongue. For $\Gamma = 0$, the tongue would be defined by two straight lines that reach down to $\lambda = 0$, see Fig. 3.5(a). The role of $\Gamma = \omega_0/Q$ becomes clear when visualizing H_{rot} in the region S' between the straight and rounded Arnold tongue tip; for these combinations of ω and λ, a saddle forms at the origin of the phase space, $u = v = 0$, which should not be a stable solution, see Fig. 3.5(b). However, as we see in eqns (3.2) and (3.3), Γ provides a force that directs the system toward the origin. Within S', this stabilization is strong enough to keep the resonator stable at $u = v = 0$, while above it the parametric drive overcomes the linear damping.

Here, we are mainly interested in understanding the conditions for stability of the system at equilibrium points in phase space. In Sections 4.2.3 and 5.1.1, we will extend the discussion to study fluctuations in the vicinity of these equilibrium points with the help of the quasi-potential Hamiltonian H_{rot} and the Jacobian matrix M.

3.1.4 Parametric Symmetry Breaking

In Section 3.1.1, we have seen how a weak parametric pump can modify the response of a driven harmonic resonator. Here, we study the opposite case, namely, a nonlinear parametric resonator pumped above threshold with a relatively weak external force perturbing its parametric phase states. Further reading can be found in Refs. [54, 56, 74].

Solving the full eqn (3.4), we obtain up to five different solutions at every frequency, out of which up to three are stable. Example solutions for one set of parameters are shown in Fig. 3.6(a), where each of the lines corresponds to a single solution. The two high-amplitude branches, which previously were degenerate in amplitude, are now slightly shifted from each other and become visible as individual lines. We also note that all solutions are displaced away from $X = 0$, and that a sweep from left to right (low to high frequency) can only sample one of the large-amplitude branches. Note that due to the interplay between the parametric pump and the external force, we observe a significant change in the bifurcation topology, that is, the position of bifurcation points in parameter space.

The modified bifurcation topology leads to new hysteretic phenomena. Performing a sweep across the solutions in Fig. 3.6(a) from low to high frequencies results in a response similar to that of the standard Duffing resonator. When sweeping in the opposite direction, we sample the lowest solution until we hit the bifurcation point close to $\omega \approx 101$. Unlike the standard Duffing, here the system has two high-amplitude solutions, out of which it selects the lower one. Continuing with the sweep, this solution becomes unstable at the bifurcation point marked as ω^*, forcing the system to jump a second time. While the amplitude difference before and after this jump may be small, the phase jump is approximately π [74] and can be used for sensing applications [56, 94].

The schematic in Fig. 3.6(b) illuminates the mechanism of parametric symmetry breaking. Depending on the phase 2θ of the external force relative to that of the parametric drive, ψ, the force shifts both phase states in one particular direction in

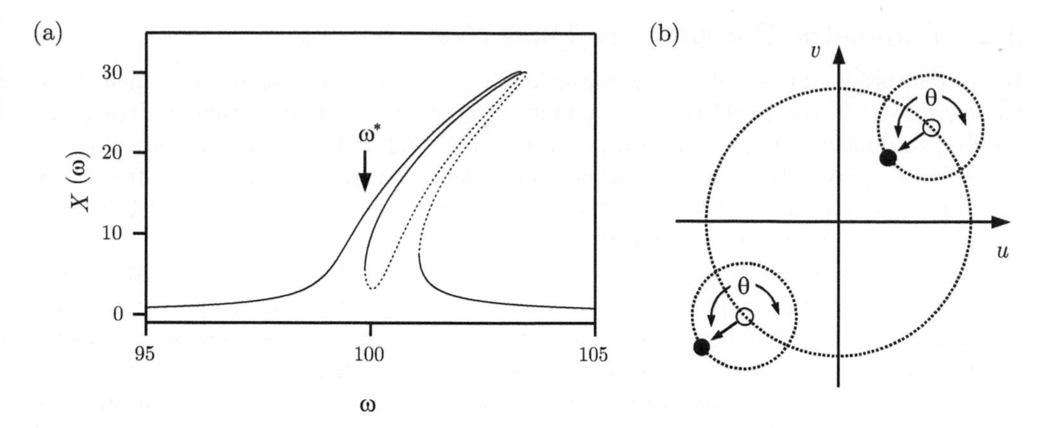

Fig. 3.6 (a) Parametric symmetry breaking obtained by solving eqn (3.4) for $\omega_0 = 100$, $m = 1$, $\Gamma = 1$, $\beta = 1$, $\eta = 0.002$, $\lambda = 1.5\lambda_{\text{th}}$, $\psi = \pi/2$, $F_0 = 800$, and $\theta = 0$. Solid and dotted lines correspond to stable and unstable solutions, respectively. A bifurcation point exists wherever a stable and an unstable solution merge. (b) The small external force shifts the phase states away from their unperturbed positions (empty circles) to new coordinates (large black dots). Depending on θ, the effect of the force can break the amplitude degeneracy and/or the phase symmetry.

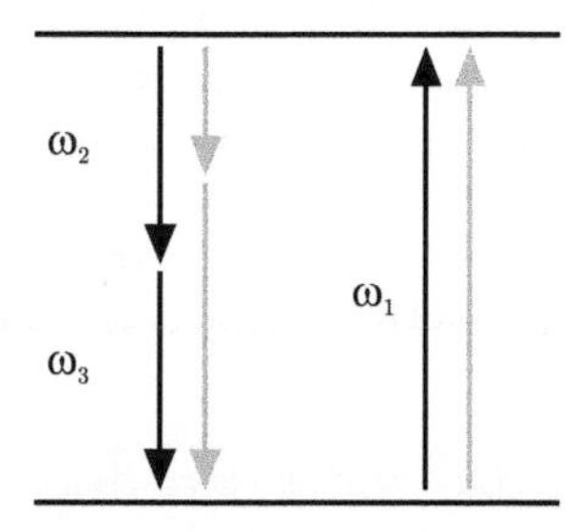

Fig. 3.7 Schematic illustration of three-wave mixing. Black and gray arrows indicate degenerate and non-degenerate mixing processes, respectively. The arrows' lengths correspond to their frequencies ω_i whose sum must be zero for a resonant effect due to energy conservation.

the phase space spanned by u and v.[1] In general, the amplitudes of the phase states are shifted in opposite directions and their phase symmetry is broken. Note, for instance, that the large black dots in Fig. 3.6 are not symmetric around the origin and that the phase separating them is no more exactly π. As long as the effect of the external force is small, however, the phase shift is not significant and we can still identify the new solutions as approximate phase states.

[1] Keep in mind that we need to compare ψ to 2θ, not θ, because of the periodicity doubling.

3.2 Parametric Pumping via Three-Wave Mixing

In our discussion of the Duffing resonator in Chapter 2, we found that a nonlinearity of the form $\beta_2 x^2$ in the EOM can be accounted for by an effective frequency-dependent Duffing nonlinearity in the slow-flow picture, cf. eqn (2.35). This useful relationship still holds for the parametric resonator as far as the dependence of the resonance frequency on amplitude is concerned. We can treat quadratic nonlinearities in eqn (3.1) with a normalized and frequency-dependent β.

There is, however, a second effect caused by the quadratic nonlinearity that we now address. This effect is related to **wave mixing** which we mentioned in Sections 2.1.1 and 2.2. Remember that nonlinear forces introduce several time-dependent terms that mix and generate forces at the sums and differences of their respective frequencies. Concretely, for a cubic potential with $n = 3$ (corresponding a quadratic nonlinearity β_2), we can resonantly excite the system if oscillation tones are present at ω_1 and ω_2 that fullfil $\omega_1 \pm \omega_2 = \pm\omega_0$, see Fig. 3.7. In the special case that $\omega_2 = \omega_0$, the condition simplifies to $\omega_1 = 2\omega_0$ and we speak of **degenerate three-wave mixing**. Note that the oscillation at ω_1 alone does not lead to resonant driving of the mode at ω_0; it only becomes effective through the downconversion effect caused by the mixing with ω_0 itself. Therefore, before a stable oscillation at ω_0 can be driven by three-wave mixing, a finite oscillation at that frequency is already required. This small starter oscillation can be provided by a second resonant drive (at $\omega_1 - \omega_0 = \omega_0$) or by thermal fluctuations, as we will introduce in Chapter 4. The force at $\omega_1 = 2\omega_0$ then amplifies this initial oscillation. The fact that degenerate three-wave mixing can be understood as an amplification process reminds us of our discussion in Section 1.4 and of the parametric amplification in Section 3.1.1.

Let us consider degenerate three-wave mixing from a second point of view; the nonlinearity leads to a change of the effective spring constant as a function of displacement. Think of the nonlinear force as

$$\beta_2 x^2 = \beta_2 x \cdot x = \tilde{\beta}_2 x \,, \tag{3.24}$$

where $\tilde{\beta}_2 = \beta_2 x(t)$ is an effective spring constant that changes over time along with the displacement $x(t)$. Over one full cycle of $x(t)$, the spring constant contribution of $\tilde{\beta}_2$ also makes one cycle, so we end up with a parametric modulation that is proportional to β_2.

To calculate the response of the resonator to degenerate three-wave mixing at $\omega \approx \omega_0$, we start with the EOM

$$\ddot{x} + \omega_0^2 x + \beta_2 x^2 + \Gamma\dot{x} = \frac{F_1}{m}\cos(2\omega t + \phi) \,, \tag{3.25}$$

where a force term F_1 is included to drive the system off-resonantly. The second-order slow-flow equations for this case are

$$\dot{u} = -\frac{u\Gamma}{2} - \frac{3\left(u^2 v + v^3\right)\beta}{8\omega} + \frac{v\left(\omega^2 - \omega_0^2\right)}{2\omega} + \frac{F_1\beta_2}{6m\omega^3}\left(u\sin\phi - v\cos\phi\right) \,, \tag{3.26}$$

$$\dot{v} = -\frac{v\Gamma}{2} + \frac{3\left(uv^2 + u^3\right)\beta}{8\omega} - \frac{u\left(\omega^2 - \omega_0^2\right)}{2\omega} - \frac{F_1\beta_2}{6m\omega^3}\left(u\cos\phi + v\sin\phi\right) \,. \tag{3.27}$$

For the case $\beta_3 = 0$, the cubic potential term leads to an effective nonlinearity $\beta = -\frac{10}{9}\frac{\beta_2^2}{\omega^2}$. Furthermore, we note a similarity between the terms $\propto F_1$ in eqns (3.26) and (3.27) and those $\propto \lambda$ in eqns (3.2) and (3.3). Degenerate three-wave mixing generates an effective parametric pump through the combination of the drive at 2ω and the nonlinearity β_2.

Motivated by this finding, we return to the full EOM including all terms we have discussed so far,

$$\ddot{x} + \omega_0^2 \left[1 - \lambda \cos\left(2\omega t + \psi\right)\right] x + \beta_3 x^3 + \beta_2 x^2 + \Gamma \dot{x} + \eta x^2 \dot{x} = \frac{F_0}{m} \cos\left(\omega t + \theta\right) + \frac{F_1}{m} \cos\left(2\omega t + \phi\right) . \tag{3.28}$$

We understand now that the second-order slow-flow equations for this case are equivalent to eqns (3.2) and (3.3) after replacing $\beta_3 \Rightarrow \beta$ as in eqn (2.35) and substituting

$$\frac{\lambda\omega_0^2}{4\omega}\left(u \sin\psi - v\cos\psi\right) \Rightarrow u\left(\frac{\lambda\omega_0^2}{4\omega}\sin\psi + \frac{F_1\beta_2}{6m\omega^3}\sin\phi\right) - v\left(\frac{\lambda\omega_0^2}{4\omega}\cos\psi + \frac{F_1\beta_2}{6m\omega^3}\cos\phi\right), \tag{3.29}$$

$$\frac{\lambda\omega_0^2}{4\omega}\left(u \sin\psi + v\cos\psi\right) \Rightarrow v\left(\frac{\lambda\omega_0^2}{4\omega}\sin\psi + \frac{F_1\beta_2}{6m\omega^3}\sin\phi\right) + u\left(\frac{\lambda\omega_0^2}{4\omega}\cos\psi + \frac{F_1\beta_2}{6m\omega^3}\cos\phi\right), \tag{3.30}$$

where eqn (3.29) and eqn (3.30) refer to the modifications to be made in eqns (3.2) and (3.3), respectively. For the case that $\phi = \psi$, we can simply define an effective parametric pumping strength

$$\lambda_{\text{eff}} = \lambda + \frac{2F_1\beta_2}{3m\omega_0^2\omega^2}. \tag{3.31}$$

The important message we take from this equation is that the presence of a quadratic nonlinearity β_2 in the restoring force allows to implement a parametric pump via an external driving force F_1. In the following, we will describe our systems with a single effective Duffing nonlinearity β and a single parametric pump λ, but will keep in mind that both can be generated, or modified, by a quadratic term β_2.

Chapter summary

- In Chapter 3, we include parametric modulation (pumping) of the potential term of the driven and damped nonlinear oscillator.
- Using the averaging method, we find the slow-flow equations of the parametrically driven nonlinear resonator, cf. eqns (3.2) and (3.3). The parametric modulation depth λ acts as an effective damping term, cf. eqn (3.8). There exists a **parametric threshold** λ_{th} above which the effective damping of the system becomes negative. The stability diagram of the damped harmonic oscillator in a space spanned by λ and ω is called the **Arnold tongue**, cf. Fig. 3.2 and eqn (3.11).
- For $\lambda < \lambda_{\text{th}}$, the effect of the parametric pump is to change the effective Q and to amplify external forces, cf. Fig. 3.1 and eqn (3.10) and (3.10). For $\lambda > \lambda_{\text{th}}$, the linear resonator becomes unstable and the system passes a **time-translation symmetry breaking bifurcation** that leads it to one out of two stable large-amplitude solutions, cf. Fig. 3.3. These solutions are referred to as **phase states** in this book.
- Revisiting methods from Chapter 1, we can use the characteristic equation to study the stability of different equilibrium points of the system, cf. eqn (3.21). Furthermore, we can define an **effective rotating-frame Hamiltonian** for the driven, nonlinear system, cf. eqn (3.22). This Hamiltonian leads us to a **quasi-potential picture** that can sometimes be useful to understand the system dynamics.
- The symmetry of the parametric phase states in the rotating phase space can be broken by a small external force, cf. Fig. 3.6. Such parametric symmetry breaking generates shifted phase states in phase space and additional solutions in a frequency sweep.
- We have seen in Chapter 2 how the nonlinear coefficient β_2 belonging to a quadratic restoring force can be integrated into an effective cubic term β. Here, we find that in concert with an external force close to $2\omega_0$, the quadratic nonlinearity can lead to an effective parametric modulation via **three-wave mixing**, cf. eqns (3.29) to (3.31). For many experimental systems, this is how parametric pumping is achieved.

Exercises

Check questions:

(a) Explain in a simple way (or in several ways) why a parametric pump leads to phase-dependent amplification.

(b) Why can we neglect nonlinearities when determining the boundaries of the Arnold tongue?

(c) Why does a nonlinear potential limit the amplitude of the parametric oscillator when pumping above the threshold?

(d) In a frequency sweep, what is the difference between the Duffing oscillator with an external force and the parametric oscillator?

(e) What is the role of damping in a parametric oscillator? What is different in the case $\Gamma = 0$?

(f) In what sense is the picture of a quasi-potential picture useful? Where does it fail?

(g) Why can a parametric pump be generated by the combination of an external drive and a quadratic nonlinearity?

Tasks:

3.1 Use the code **Python Example 2** to test parametric pumping with an external drive F_1, cf. eqn (3.31). Use $\omega = \omega_0 = 1$, $\Gamma = 0.001$, $\beta_2 = 0.01$, $\psi = \phi = -\pi/2$, $x_{\text{in}} = 0.1$, and a sampling time step of 0.01 as default values (all other values equal to zero). Compare ringdowns without and with a parametric pump close to (or at) the threshold, $\lambda \leq \lambda_{\text{th}}$. Then set $\lambda = 0$ and calculate what value you need for F_1 to reach the parametric threshold with three-wave mixing. Compare the ringdown with F_1 below and above this value. What are the similarities to the standard parametric pump, and what are the differences? Use the command **axes.set_xlim(...)** to zoom in on your result and study the short-scale behavior of the oscillation.

3.2 Open **Python Example 3** and study the driven harmonic oscillator with a parametric drive below threshold, cf. eqn (3.8). Use $\omega = \omega_0 = 1$ and $\Gamma = 0.01$ as default values. Start from the initial condition **x0=0** and drive the oscillator with an external force. Activate a parametric pump and observe the phase-dependent parametric amplification for values $\lambda \leq \lambda_{\text{th}}$.

3.3 Repeat the previous exercise with a frequency sweep. Try in particular to reduce the amplitude by selecting the correct phase. Why is the observed amplification not uniform as a function of the driving frequency ω?

3.4 Observe the behavior of the system in the presence of a parametric drive above threshold. Use $F_0 = 0$, $\omega = \omega_0 = 1$, $\Gamma = 0.01$, and $\beta = 0.1$ as default values and try various initial conditions. How many stable solutions do you find?

3.5 Perform frequency sweeps in both directions with the same parameters as above. Can you see a hysteresis as in Fig. 3.2(b)? What is the difference from the result obtained with the Duffing oscillator in a frequency sweep?

3.6 Test parametric symmetry breaking as shown in Fig. 3.6. Can you see the double hysteresis in a frequency sweep? What happens for an increasing external force F_0? For very large F_0, what happens to the number of stable solutions?

3.7 Test parametric symmetry breaking in frequency sweeps with various phases of the external force (θ). Can you explain the different results?

3.8 In a parametric oscillator with symmetry-breaking force and $\theta = 0$, extract the jump frequency ω^* as a function of F_0. Can you think of an application for this effect?

4

Dissipation and Force Fluctuations

In the previous chapters, we included in our EOM a damping coefficient Γ that describes the rate at which the resonator loses energy, without contemplating where this energy is going to. In this chapter, we address the role of the environment and the fluctuations it creates within our system.

4.1 The Role of Force Noise

A realistic model of a resonator includes an environment that accepts the energy leaking out, and the coupling between the resonator and the environment should be allowed to transport energy both ways, see Fig. 4.1. This environment is assumed to have a well-defined temperature T, meaning that all its degrees of freedom have the same average total energy, cf. eqn (1.1). In the absence of external driving and parametric pumping and after a sufficiently long time, any resonator is expected to be in **thermal equilibrium** with the environment it is coupled to. The resonator then has a total energy given by

$$E_{\rm eq} \equiv \langle H \rangle = \frac{1}{2} m\omega_0^2 \langle x^2 \rangle + \frac{1}{2} m \langle \dot{x}^2 \rangle = m\omega_0^2 \langle x^2 \rangle = k_B T\,, \tag{4.1}$$

where $k_B = 1.38 \times 10^{-23}\,\mathrm{J/K}$ is the Boltzmann constant, $\langle ... \rangle$ denotes average over time, and we make use of the fact that potential and kinetic energy are on average equal over one oscillation cycle. We set $\langle x \rangle = 0$ whenever not explicitly mentioned, such that $\langle x^2 \rangle$ becomes equal to the variance $\sigma_x^2 \equiv \langle x^2 \rangle - \langle x \rangle^2$, and likewise for $\dot{x}$. Equation (4.1) is a way of writing the **equipartition theorem** [98].

How can we fullfil the equipartition theorem in our formal description of the resonator? Previously, we only considered loss of energy from the resonator to the environment through a dissipation term $\propto \Gamma$, which leads to $\langle H \rangle = 0$ in the long-time limit. We now need to add some driving term to achieve a thermal balance. This driving term should not have any preferred frequency or phase, because we cannot identify any reason why the environment should generally possess such features. A realistic model of such a driving term rests on stochastic kicks acting on the resonator. The kicks are delivered by the ensemble of microscopic degrees of freedom in the environment. They can arrive at any moment and are not correlated, such that no preferred frequency component is expected. Consequently, we model the influence of the environment on the resonator by a force term $\xi(t)$ whose amplitude and phase are fluctuating over time and cover all frequency components equally. We will refer to this as **thermal force noise**.

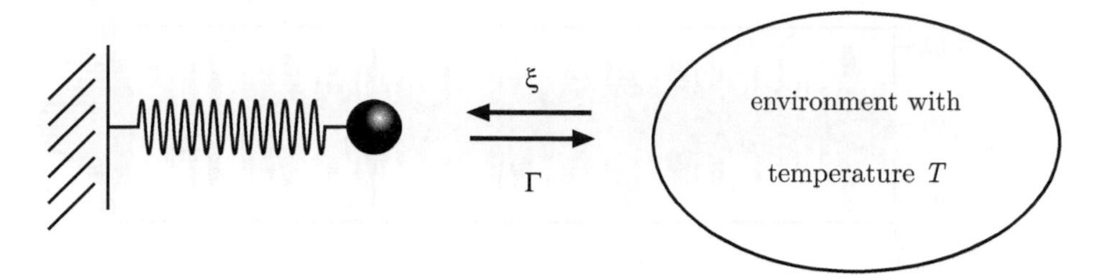

Fig. 4.1 The resonator is coupled to an environment with a mean energy density expressed by a temperature T. The damping rate Γ describes the energy loss from the resonator to the environment. At the same time, the environment acts on the resonator through a fluctuating force ξ. The two phenomenological parameters Γ and ξ are manifestations of the same coupling.

Starting again with a linear resonator, we write the EOM

$$\ddot{x} + \omega_0^2 x + \Gamma \dot{x} = \frac{F_0}{m} \cos(\omega t + \theta) + \frac{\xi}{m}. \tag{4.2}$$

The addition of a fluctuating force turns eqn (4.2) into a stochastic differential equation. Writing such a system in the integral form of eqn (1.28) leads to a so-called **Îto process**

$$\mathbf{x}(t) = \Phi(t,0)\mathbf{x}(0) + \int_0^t \Phi(t,t') \frac{1}{m} \begin{pmatrix} 0 \\ F_0 \cos(\omega t') + \xi(t') \end{pmatrix} dt'. \tag{4.3}$$

To be able to perform the integral, the force term is assumed to be the time derivative of a random walk, $\xi(t) = \dot{B}(t)$. The random walk $B(t)$ is an example of a **Wiener process** that is smooth at arbitrarily short timescales, which implies that ξ acts at arbitrarily high frequencies.[1] Formally, our only assumptions about the stochastic force ξ are

$$\begin{aligned} \langle \xi(t) \rangle &= 0\,, \\ \langle \xi(t)\xi(t+\Delta_t) \rangle &= \varsigma_D^2 \delta(\Delta_t)\,, \end{aligned} \tag{4.4}$$

where $\delta(...)$ denotes the delta function, Δ_t is a time delay, and ς_D is a normalization factor that we define in the following. Equation (4.4) expresses that the values of ξ at two times t and $t + \Delta_t$ are uncorrelated unless $\Delta_t = 0$. The Fourier transform of the autocovariance in eqn (4.4) must be completely *white*, that is, have the same intensity at all frequencies up to ∞ (or at least higher than any frequency we are interested in),

$$\begin{aligned} S_F(\omega/2\pi) &= \int_0^{t_S} \langle \xi(t)\xi(t+\Delta_t) \rangle\, e^{-i\omega\Delta_t} d\Delta_t \\ &= \varsigma_D^2 \int_0^{t_S} \delta(\Delta_t) e^{-i\omega\Delta_t} d\Delta_t = \varsigma_D^2\,, \end{aligned} \tag{4.5}$$

[1] For all practical purposes, a finite time resolution suffices such that the formalism can be applied to realistic situations.

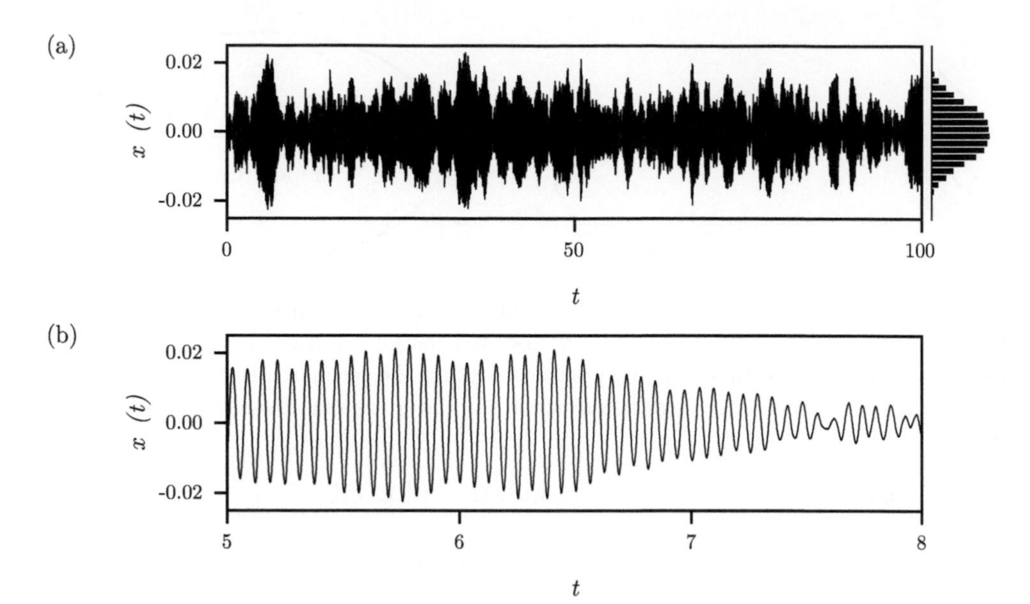

Fig. 4.2 Numerical simulation of a resonator driven by force noise with $\omega_0 = 100$, $Q = 100$, $m = 1$ and $\varsigma_D = 1$, cf. eqn (4.4). (a) Time trace sampled with $t_{\text{samp}} = 0.001$. The sampled amplitudes are collected in a histogram (right), indicating a Gaussian distribution. (b) Zoom of the data in (a). Oscillations are smooth on short timescales. The correlation time for this example is $\tau_0 \equiv 2/\Gamma = 2$.

where we consider a noise process whose time runs from zero to a final sampling time, $0 < t < t_S$, and we use the sifting property of the Dirac delta function. The Fourier transform of an autocovariance is a power spectral density (PSD), that is, $S_F = \varsigma_D^2$ has units of N^2/Hz and quantifies the squared force acting on the resonator within a frequency range of 1 Hz. We have thus defined a force noise ξ that drives our resonator with a white PSD S_F.[2]

4.1.1 Response to Force Noise as a Function of Time

The behavior of a resonator driven by a fluctuating force ξ cannot, by construction, be predicted precisely. This means that we cannot analytically use integration to calculate $x(t)$ at arbitrary times, as we did in eqn (1.28). Nevertheless, we can use iterative numerical simulations to approximate the behavior with a sampling rate $1/t_{\text{samp}}$, where $t_{\text{samp}} \ll T_0 = 2\pi/\omega_0$. Such a *time trace* (or sampling path) for a linear resonator is simulated in Fig. 4.2. On long timescales, the amplitude and phase of the response

[2] We emphasize that we regard S_F as a function of temporal frequency $\nu = \omega/2\pi$ unless explicitly stated otherwise. When considered as a function of angular frequency ω, the value of S_F has to be divided by 2π in order to yield the same variance over a given frequency interval:

$$\int_{\nu_1}^{\nu_2} S_F(\nu)d\nu = \int_{\omega_1}^{\omega_2} S_F(\omega)d\omega = \int_{\omega_1}^{\omega_2} \frac{S_F(\nu)}{2\pi}d\omega\,. \tag{4.6}$$

In order to keep the notation simple, we use $S_F(\omega/2\pi)$ instead of $S_F(\nu)$ in the following.

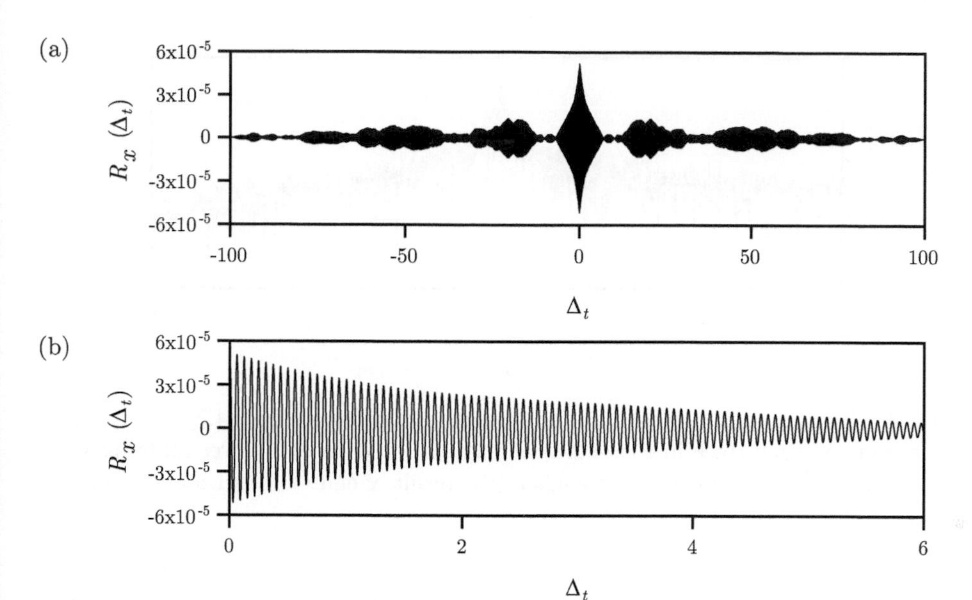

Fig. 4.3 (a) Autocovariance of the data set in Fig. 4.2, see eqn (4.7). Small peaks at $\Delta_t = \pm 20$ and ± 50 are random correlations caused by the finite measurement time, $t_S = 100$. (b) A close-up of (a) near $\Delta_t = 0$ shows oscillations between positive and negative correlation values alongside an overall decay on a timescale of $\tau_0 = 2$.

appear entirely uncorrelated, see Fig. 4.2(a). This is what we naively expect from the statistical properties of ξ in eqn (4.4). However, we observe that the oscillations are approximately smooth on timescales shorter than $\tau_0 \equiv 2/\Gamma$, see Fig. 4.2(b).[3] This might be surprising until we remember that the resonator is a band-pass filter which amplifies force components close to ω_0 and suppresses off-resonant contributions. Consequently, the trace in Fig. 4.2(b) is dominated by near-resonant, smooth oscillations on short timescales. In the long-time limit, the stochastic kicks from the environment lead to a random walk process balanced by the damping. The balance generates a Gaussian distribution function in x, as shown in the histogram on the right axis of Fig. 4.2(a).

The statistical properties of a time trace can be visualized by calculating its autocovariance $R_x(\Delta_t) = \langle x(t)x(t+\Delta_t)\rangle$ with the formula

$$R_x(\Delta_t) = \lim_{t_s \to \infty} \frac{1}{t_S} \int_0^{t_S} x(t)x(t+\Delta_t)dt\,. \tag{4.7}$$

The autocovariance R_x of the data in Fig. 4.2 is shown in Fig. 4.3(a) for $-t_S \leq \Delta_t \leq t_S$. It is symmetric around $\Delta_t = 0$, where it reaches its maximum value $R_x(\Delta_t = 0) = \langle x^2\rangle = \sigma_x^2$. This means that we can retrieve the variance of the signal by setting $\Delta_t = 0$ in eqn (4.7). In this example, R_x does not decrease monotonically with $|\Delta_t|$

[3] τ_0 is identical to the exponential decay time of the amplitude ringdown and to twice that of the energy ringdown, cf. eqn (1.6) and Fig. 1.4(a).

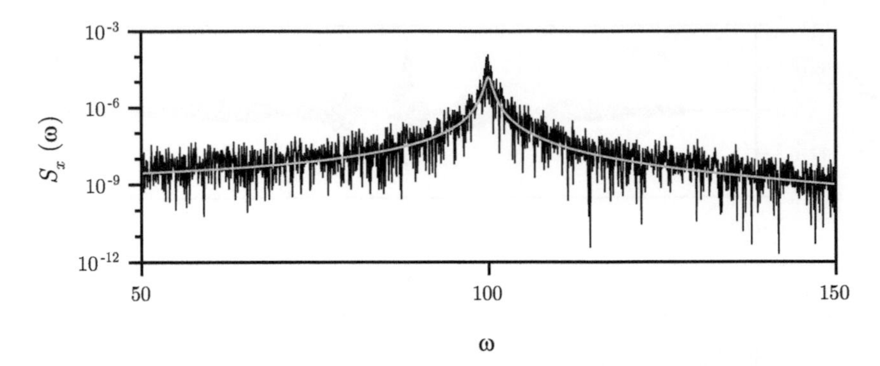

Fig. 4.4 Displacement PSD (black line) of the sample path from Fig. 4.2, cf. eqn (4.8). The variance of the time trace in Fig. 4.2(a) and the integration of S_x over all frequencies both yield 5.2×10^{-5}. A gray line corresponds to the result of eqn (4.10) with angular units, $S_x(\omega) = S_F(\omega)\chi^2$.

but exhibits small peaks. These peaks are a consequence of the finite measurement time t_S; amplitude fluctuations in the time trace in Fig. 4.2(a) do not fully average out over this time span and lead to random correlations in R_x. For $t_S \to \infty$, the decay of $R_x(t_{\text{samp}})$ is expected to follow the exponential form $e^{-\Gamma|\Delta_t|/2}$, similar to the amplitude decay in the ringdown trace in Fig. 1.4(a).

In Fig. 4.3(b), we observe that R_x oscillates on small timescales of Δ_t. The maxima in these oscillations correspond to time delays Δ_t that are equal to integer multiples of the natural period $2\pi/\omega_0$ of $x(t)$, which is a distinct signature of the phase coherence in an underdamped system. The phase coherence is lost on a timescale of τ_0. In the next section, we will analyze what role this coherence plays for the displacement PSD S_x.

4.1.2 Spectral Response to Force Noise

The **Wiener–Khinchin theorem** states that the PSD and the autocovariance of a quantity are connected through Fourier transformations, a relation that was applied to obtain eqn (4.5). We can therefore calculate the displacement PSD S_x from the Fourier transform of R_x,

$$S_x(\omega/2\pi) = \int_{-t_S}^{t_S} R_x(\Delta_t) e^{-i\omega\Delta_t} d\Delta_t \,. \tag{4.8}$$

The result of eqn (4.8) for the sample path in our example is shown in Fig. 4.4 as a black line graph. As for $S_F(\omega) = S_F(\omega/2\pi)/2\pi$, we can use the relation $S_x(\omega) = S_x(\omega/2\pi)/2\pi$ to convert between a PSD as a function of temporal and angular frequency.

Let us compare the spectral data to the analytical model of a harmonic resonator. From our treatment in Section 1.2, we know that the amplitude response of a driven and damped resonator to a force F_0 at frequency ω is given by

$$x_0^2(\omega) = \frac{F_0^2}{m^2\left(\omega_0^2 - \omega^2\right)^2 + m^2\omega^2\Gamma^2} = F_0^2\chi^2(\omega)\,, \tag{4.9}$$

where we have defined a susceptibility function χ.[4] If we imagine a force noise PSD that acts at all frequencies simultaneously, we conclude that the resonator will be driven at all frequencies with a weighted response

$$S_x = S_F\chi^2\,. \tag{4.10}$$

This analytical prediction for S_x is drawn as a gray line in Fig. 4.4. The PSD has a peak at ω_0 and a FWHM of $\Delta\omega = \Gamma = 2/\tau_0$, as expected. These are the spectral signatures of the oscillations and their decay with increasing Δ_t seen in Fig. 4.3(b).

Since S_x corresponds to the time-averaged displacement power per unit frequency $\omega/2\pi$,[5] the integration of S_x over all frequencies corresponds to the total variance of the fluctuations,

$$\langle x^2\rangle = \int_{-\infty}^{\infty} S_x(\omega)d\omega = \int_{-\infty}^{\infty} S_F(\omega)\chi^2 d\omega\,. \tag{4.11}$$

This important equation is known as **Parsevals's theorem.**

4.1.3 Discrete Signals

The results in this chapter are presented in the form of continuous signals and their integration over time. However, we should be aware that any real measurement produces discrete data points with a sampling interval t_{samp}. In this case, the continuous formula for the autocovariance in eqn (4.7) will be replaced by

$$R_x(\Delta_t) = \lim_{N\to\infty}\frac{1}{N}\sum_{n=1}^{N} x(t)x(t+\Delta_t)\,, \tag{4.12}$$

with $\Delta_t = nt_{\mathrm{samp}}$. The Wiener–Khinchin theorem in the discrete case becomes

$$S_x(\omega/2\pi) = \sum_{n=0}^{N} R_x(\Delta_t)e^{-i\omega\Delta_t}t_{\mathrm{samp}}\,. \tag{4.13}$$

The same result can be obtained from

$$S_x(\omega/2\pi) = \lim_{N\to\infty}\frac{t_{\mathrm{samp}}}{N}\left|\sum_{n=0}^{N} x_n e^{-i\omega n t_{\mathrm{samp}}}\right|^2\,, \tag{4.14}$$

where $x_n = x(nt_{\mathrm{samp}})$ is the nth value of $x(t)$.

For a discrete signal, the limits of the two-sided PSD are derived from the condition that at least two sampling points per oscillation period are needed to avoid aliasing, see

[4] The value of the susceptibility is the same for temporal and angular frequencies.

[5] The term *power* is used in the loose meaning of a quantity that is proportional to energy, not in the precise physical definition of energy per unit time.

(a)

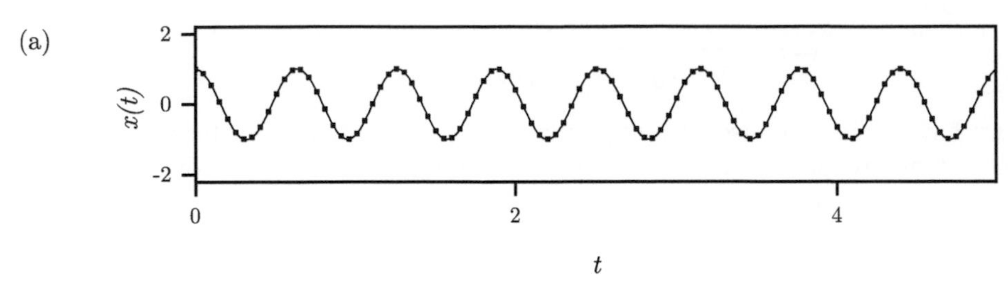

(b)

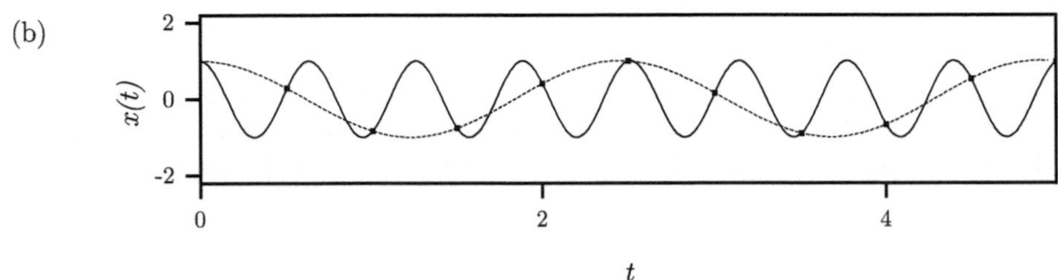

Fig. 4.5 Sine wave signal with a frequency of $\omega/2\pi = 5/\pi$ (black line) sampled in discrete time steps (black squares). In (a), the time step is $t_{\mathrm{samp}} = 0.05 < \pi/\omega$, yielding more than two points per oscillation period. This allows for a faithful estimation of the frequency. In (b), we have $t_{\mathrm{samp}} = 0.5 > \pi/\omega$, and therefore less than two points per period. In this case, the frequency cannot be estimated correctly. The (wrong) oscillation signal estimated from the sampled points is shown as a dashed line.

Fig. 4.5. This condition is captured in the Nyquist frequency $\omega_N = \frac{2\pi}{2t_{\mathrm{samp}}} = \pi/t_{\mathrm{samp}}$ as a boundary for the spectrum. The discrete double-sided PSD is formed by N points spread between $\pm\omega_N$, so its resolution is

$$\delta\omega \equiv |\omega_n - \omega_{n-1}| = \frac{2\omega_N}{N} = \frac{2\pi}{t_S}\,. \tag{4.15}$$

This means that two neighboring points of the PSD differ by exactly one full oscillation over the total measurement time t_S. Finally, Parseval's theorem takes the form

$$\langle x^2 \rangle = \delta\omega \sum_{-\omega_N}^{\omega_N} S_x(\omega)\,. \tag{4.16}$$

4.2 The Fluctuation–Dissipation Theorem

The motivation behind defining a stochastic force ξ was to achieve equilibrium between the resonator energy and the environment in the absence of other driving sources. In equilibrium, as is defined in the equipartition theorem in eqn (4.1), the rate of energy lost through the channel Γ is equal to that supplied by ξ, see Fig. 4.1. The value of $S_F = \varsigma_D^2$ must therefore be related to Γ or, in other words, the force fluctuations must be related to the energy dissipation. This notion leads to the **fluctuation–dissipation theorem** (FDT).

4.2.1 The FDT in the Nonrotating Frame

We begin by applying eqn (1.6) to transform eqn (4.2) into an expanding frame of reference and obtain the two coupled first-order differential equations

$$\dot{\mathbf{y}} = \begin{pmatrix} 0 & 1 \\ -\omega_\Gamma^2 & 0 \end{pmatrix} \mathbf{y} + \frac{e^{(\Gamma t/2)}}{m} \begin{pmatrix} 0 \\ F_0 \cos(\omega t) + \xi(t) \end{pmatrix} \tag{4.17}$$

with $\omega_\Gamma^2 = \omega_0^2 - \frac{\Gamma^2}{4}$ and where we use the vector notation

$$\mathbf{y} = \begin{pmatrix} y \\ \dot{y} \end{pmatrix} . \tag{4.18}$$

For this system of equations, we can choose the basis solutions $y_1 = \cos\Big(\omega_\Gamma t\Big)$ and $y_2 = \sin\Big(\omega_\Gamma t\Big)$, yielding the Wronskian matrix

$$W(t) = \begin{pmatrix} \cos\Big(\omega_\Gamma t\Big) & \sin\Big(\omega_\Gamma t\Big) \\ -\omega_\Gamma \sin\Big(\omega_\Gamma t\Big) & \omega_\Gamma \cos\Big(\omega_\Gamma t\Big) \end{pmatrix} , \tag{4.19}$$

and the state transition matrix

$$\Phi(t,t') = \begin{pmatrix} \cos\Big(\omega_\Gamma (t-t')\Big) & \frac{1}{\omega_\Gamma} \sin\Big(\omega_\Gamma (t-t')\Big) \\ -\omega_\Gamma \sin\Big(\omega_\Gamma (t-t')\Big) & \cos\Big(\omega_\Gamma (t-t')\Big) \end{pmatrix} \equiv \begin{pmatrix} \Phi_1 \\ \Phi_2 \end{pmatrix} , \tag{4.20}$$

where we have defined the row vectors $\Phi_{1,2}$ for later use. With initial conditions $\mathbf{y}_0 = \mathbf{y}(0)$, we can now evolve $\mathbf{y}$ as in eqn (1.28),

$$\mathbf{y}(t) = \Phi(t,0)\mathbf{y}_0 + \int_0^t \frac{e^{(\Gamma \tau/2)}}{m} \Phi(t,t') \begin{pmatrix} 0 \\ F_0 \cos(\omega t') + \xi(t') \end{pmatrix} dt' . \tag{4.21}$$

The problem we are facing here is that, given that ξ is uncorrelated over any finite timescale, its integration will produce no average contribution to $\mathbf{y}(t)$, while the instantaneous contributions are by definition unpredictable. The expectation value of $\langle y \rangle$ from eqn (4.21) is therefore the same as that for $\xi = 0$. This is not surprising, because the original motivation behind introducing the notion of a fluctuating force noise was to fulfill a condition for $\langle x^2 \rangle$, not $\langle x \rangle$, see eqn (4.1). To calculate the effect of ξ on $\langle \mathbf{y}^2 \rangle$, we square eqn (4.21) and find

$$\begin{aligned} \langle \mathbf{y}^2 \rangle =& \langle \mathbf{y} \rangle^2 + 2 \left(\Phi_1(t,0)\mathbf{y}_0 + \int_0^t \frac{e^{\Gamma t'/2}}{m} \Phi_1(t,t') \begin{pmatrix} 0 \\ F_0 \cos(\omega t') \end{pmatrix} dt' \right) \\ & \times \int_0^t \frac{e^{\Gamma t'/2}}{m} \Phi_1(t,t') \begin{pmatrix} 0 \\ \langle \xi(t') \rangle \end{pmatrix} dt' \\ & + \langle \int_0^t \frac{e^{\Gamma t'/2}}{m} \Phi_1(t,t') \begin{pmatrix} 0 \\ \xi(t') \end{pmatrix} dt' \int_0^t \frac{e^{\Gamma t''/2}}{m} \Phi_1(t,t'') \begin{pmatrix} 0 \\ \xi(t'') \end{pmatrix} dt'' \rangle . \end{aligned} \tag{4.22}$$

From our definition in eqn (4.4), we know that $\langle\xi\rangle = 0$ and that $\xi(t')$ and $\xi(t'')$ are uncorrelated for $t' \neq t''$. Focusing on the displacement (upper part of the vector) we can therefore simplify the equation to

$$\begin{aligned}\langle y^2\rangle &= \langle y\rangle^2 + \frac{1}{\omega_\Gamma^2}\int_0^t\int_0^t \frac{e^{\Gamma t'}}{m^2}\sin\Big(\omega_\Gamma(t-t')\Big)\sin\Big(\omega_\Gamma(t-t'')\Big)\varsigma_D^2\delta(t'-t'')dt'dt'' \\ &= \langle y\rangle^2 + \frac{\varsigma_D^2}{\omega_\Gamma^2}\frac{\Gamma^2(1-\cos\Big(2\omega_\Gamma t\Big)) - 4\omega_\Gamma^2\left(e^{\Gamma t}-1\right) + 2\Gamma\omega_\Gamma\sin\Big(2\omega_\Gamma t\Big)}{-2\left(\Gamma^3+4\Gamma\omega_\Gamma^2\right)m^2}\,. \end{aligned} \tag{4.23}$$

The stationary response to F_0 is encoded in the first term $\langle y\rangle^2$. However, we are interested in the second term, which describes the variance of the displacement fluctuations regardless of the mean value $\langle y\rangle$. Therefore, without loss of generality we utilize $\langle y\rangle = 0$ and transfer back to $x(t) = e^{-\Gamma t/2}y(t)$. Neglecting terms $\propto \Gamma^3$, our final result is

$$\langle x^2\rangle = \frac{\varsigma_D^2}{2m^2\Gamma\omega_0^2}\,. \tag{4.24}$$

Let us compare this to our original idea for introducing a force noise in to the EOM. The equipartition theorem in eqn (4.1) asserts that $m\omega_0^2\langle x^2\rangle = k_BT$. Inserting this into eqn (4.24) reveals that

$$S_F(\omega/2\pi) = \varsigma_D^2 = 2k_BTm\Gamma\,. \tag{4.25}$$

This surprisingly simple relationship between the fluctuating force PSD S_F, the damping coefficient Γ, and the temperature T is known as the FDT. We present it here for a double-sided convention, that is, for a spectral distribution from $-\infty$ to ∞. In many experimental situations, it is more convenient to use a single-sided convention that takes only positive frequencies from 0 to ∞ (or to a maximum Nyquist frequency ω_N) into account. To comply with Parseval's theorem in eqn (4.11), the factor 2 in eqns (4.24) and (4.25) is then replaced by a factor 4.

4.2.2 The FDT in the Rotating Frame

In order to study nonlinear parametric phenomena, we found it beneficial to perform a translation into a rotating frame, cf. eqn (2.22). We will therefore study how thermal force noise manifests in the slow frame.

Applying the Krylov–Bogolyubov averaging method from Section 2.1.3 to lowest order, we find that in the slow-flow equations, the stochastic force ξ is expressed in the form of two force terms Ξ_u and Ξ_v that drive u and v, respectively [99, 100]. The terms are

$$\begin{aligned}\Xi_u(t) &= -\frac{1}{T_0}\int_t^{t+T_0}\frac{\sin(\omega t')\,\xi(t')}{\omega}dt'\,, \\ \Xi_v(t) &= -\frac{1}{T_0}\int_t^{t+T_0}\frac{\cos(\omega t')\,\xi(t')}{\omega}dt'\,, \end{aligned} \tag{4.26}$$

where we remember T_0 to be the natural period of the resonator. The averaging defines an effective low-pass filter, so the force noise becomes correlated at times shorter than

T_0. However, as $T_0 \ll \tau_0$, this correlation between averaged force noises does not influence the behavior of our system, as τ_0 itself imposes a much stricter filter. We therefore neglect the short-term correlation of the averaged force noise and write

$$\langle \Xi_i(t)\Xi_j(t')\rangle = \frac{\varsigma_D^2}{2\omega^2}\delta_{ij}\delta(t-t'), \tag{4.27}$$

where the delta function δ_{ij} signifies that two different stochastic force terms are mutually uncorrelated. The slow-flow equations are accordingly modified and become (for a driven, damped harmonic oscillator, cf. eqn (4.2))

$$\dot{u} = -\frac{u\Gamma}{2} + \frac{v\left(\omega^2-\omega_0^2\right)}{2\omega} + \frac{F_0\sin(\theta)}{2m\omega} + \frac{\Xi_u}{m}, \tag{4.28}$$

$$\dot{v} = -\frac{v\Gamma}{2} - \frac{u\left(\omega^2-\omega_0^2\right)}{2\omega} - \frac{F_0\cos(\theta)}{2m\omega} + \frac{\Xi_v}{m}. \tag{4.29}$$

We can integrate these equations, similar to what we have done in eqn (4.21). We are again interested in the long-time limit where the initial conditions are unimportant, and we set $F_0 = 0$ such that $\langle u\rangle = \langle v\rangle = 0$. With the definition

$$A = \begin{pmatrix} \Gamma\omega & (\omega_0^2-\omega^2) \\ -(\omega_0^2-\omega^2) & \Gamma\omega \end{pmatrix}, \tag{4.30}$$

we obtain for the squared amplitude $X^2 = u^2+v^2$ the equation

$$\begin{aligned}\langle X^2\rangle &= \left|\frac{1}{m^2}\int_0^t\int_0^t \langle e^{-\frac{A}{2\omega}(t-t')}\begin{pmatrix}\Xi_u(t')\\ \Xi_v(t')\end{pmatrix} e^{-\frac{A}{2\omega}(t-t'')}\begin{pmatrix}\Xi_u(t'')\\ \Xi_v(t'')\end{pmatrix}\rangle \mathrm{d}t'\mathrm{d}t''\right| \\ &= \left|\int_0^t \frac{\Xi_u^2+\Xi_v^2}{m^2} e^{\frac{1}{2}\Gamma(t''+t'-2t)}\cos\left(\frac{(t''-t')\left(\omega^2-\omega_0^2\right)}{2\omega}\right)\mathrm{d}t'\mathrm{d}t''\right| \\ &= \frac{2\varsigma_D^2}{2m^2\omega^2}\int e^{\frac{1}{2}\Gamma(2t'-2t)}\mathrm{d}t' = \frac{\varsigma_D^2}{m^2\Gamma\omega^2}(1-e^{-\Gamma t}).\end{aligned} \tag{4.31}$$

As the fluctuations will appear mostly on resonance, we set $\omega = \omega_0$. In the long-time limit $t \to \infty$, we arrive at

$$\varsigma_D^2 = m^2\Gamma\omega_0^2\langle X^2\rangle. \tag{4.32}$$

We see that the FDT for the averaged amplitude X differs from the non-rotating picture by a factor 2, cf. eqn (4.25). This is due to the fact that in the rotating frame, both quadratures (u and v) are subject to a fluctuating force, while in the nonrotating frame we defined a force term to drive x but none for $\dot{x}$. Considering that

$$E_{\mathrm{eq}} = \frac{1}{2}m\omega_0^2\langle u^2\rangle + \frac{1}{2}m\omega_0^2\langle v^2\rangle = \frac{1}{2}m\omega_0^2\langle X^2\rangle = k_BT, \tag{4.33}$$

we see that eqn (4.32) is analogous to the equipartition theorem of eqn (4.25) in the rotating frame, in the sense that the energy distributes equally between both oscillation quadratures. In the absence of an external force, we assume $\langle u\rangle = \langle v\rangle = 0$, such that we can insert the variances $\sigma_u^2 \equiv \langle u^2\rangle - \langle u\rangle^2 = \langle u^2\rangle$ and $\sigma_v^2 \equiv \langle v^2\rangle - \langle v\rangle^2 = \langle v^2\rangle$ into eqn (4.33). However, it is important to remember that $\langle X\rangle \neq 0$, such that in general $\sigma_X^2 \neq \langle X^2\rangle$.

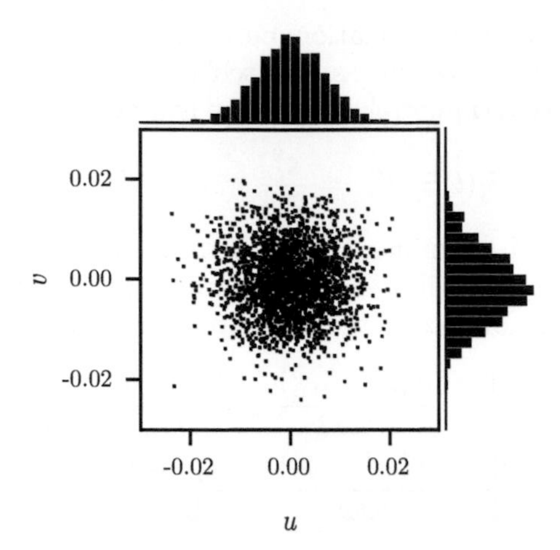

Fig. 4.6 Numerical simulation of a resonator driven by force noise with $\omega_0 = 100$, $Q = 100$, $m = 1$, $F_0 = 0$, and $\varsigma_D = 1$. The white force noise leads to a Gaussian amplitude distribution in both quadratures, as shown by the histograms. The evolution was simulated over a time of 5000 and the frame rotates at $\omega = \omega_0$.

4.2.3 Effect of Force Noise in the Rotating Frame

It is worth having a closer look at the behavior of the thermally driven oscillator in the rotating frame. We start by noting that for $\Xi_{u,v} \to 0$, eqns (4.28) and (4.29) yield the equilibrium points u_{eq} and v_{eq} of the driven, damped oscillator. In the rotating-frame stability analysis that we presented in Section 3.1.3, a stable equilibrium point is an attractor that is invariant under small perturbations δu and δv. Indeed, the system fluctuates around average coordinates $\langle u \rangle$ and $\langle v \rangle$ with a random phase. In the long-time limit, the histograms of u and v approach normal distributions whose mean values $\langle u \rangle$ and $\langle v \rangle$ converge again toward u_{eq} and v_{eq}, see Fig. 4.6.

The force noise does increase the variances σ_u^2 and σ_v^2; after all, we introduced the notion of a fluctuating force precisely with the aim of accounting for the thermal energy of the resonator mode that is proportional to the displacement variance. From a comparison of eqns (4.1) and (4.33), we find $\sigma_u^2 = \sigma_v^2 = \sigma_x^2 = \sigma_{\dot{x}}^2/\omega_0^2$.

The two coordinates u and v fluctuate as a function of time under the influence of Ξ_u and Ξ_v, just as x and $\dot{x}$ fluctuate under the influence of ξ. However, we should emphasize an important difference between the coupled differential equations in eqns (4.28) and (4.29) and those in the nonrotating frame that we saw for example in eqn (1.18). When the frame rotates exactly at $\omega = \omega_0$ and for $F_0 = 0$, the slow-flow equations eqns (4.28) and (4.29) are reduced to

$$\dot{u} = -\frac{u\Gamma}{2} + \frac{\Xi_u}{m}, \tag{4.34}$$

$$\dot{v} = -\frac{v\Gamma}{2} + \frac{\Xi_v}{m}. \tag{4.35}$$

Obviously, the degrees of freedom u and v are now decoupled and respond individually to forces in their respective quadratures. There is no notion of a resonance frequency in the rotating-frame system, and both u and v are subject to simple first-order differential equations. The trajectories of the two quadratures describe random walks in the presence of a linear centripetal force given by Γ. Note that the damping force is proportional to a displacement, not to a velocity as in the nonrotating frame, and that there is no term to assume the role of a momentum. The only relevant timescale in this situation is the time constant τ_0 that characterizes the decay of the system toward the equilibrium position after it has been displaced. We will see in Chapter 5 that these observations no longer hold in a nonlinear resonator.

4.3 The Probability Distribution Approach

So far, we have considered the behavior of a resonator as a sequence of steps in time. After some time, the resonator settles into its steady state, force noise leads to fluctuations around that steady state, and by observing these fluctuations for long enough, one can make predictions as to how probable it is to find the resonator at a given amplitude and phase in the future.

There is a second approach to making such statistical predictions. Namely, we can ignore the fact that a classical resonator has one specific amplitude and phase at any given moment, and consider only the probability distribution $\rho(\mathbf{x}, t)$ that describes how probable it is that the system will be at the position x with velocity $\dot{x}$ at time t. In the absence of noise, the behavior of a classical resonator is completely deterministic and the state $\mathbf{x}(t)$ can be predicted with certainty. In this deterministic case, $\rho(\mathbf{x}, t)$ must be zero everywhere except at $\mathbf{x}(t)$. Uncertainty about the initial conditions of the system or the forces acting on it broadens $\rho(\mathbf{x}, t)$. The probability distribution and its evolution over time can be used to describe the resonator state at any given time.

4.3.1 Liouville's Theorem

The formalism for treating the evolution of $\rho(\mathbf{x}, t)$ originates from statistical differential equations in thermodynamics and chemistry [101, 102]. There, the notion of a density function arises naturally from problems that involve very large numbers of individual particles, such as molecules or atoms. The particles form a quasi-continuous cloud whose density over an infinitesimal volume is obtained by counting the particles within. When the dynamics of every particle in such a cloud obeys Hamilton's equations in the absence of driving or damping, the Liouville theorem states that the equilibrium density function does not change with time,

$$\frac{d\rho}{dt} = \frac{\partial \rho}{\partial t} + \sum_{j=1}^{N} \left(\frac{\partial \rho}{\partial x_j} \dot{x}_j + \frac{\partial \rho}{\partial p_j} \dot{p}_j \right) = 0 \,, \tag{4.36}$$

where x_j and $p_j = m\dot{x}_j$ are the position and momentum coordinates of the particle with index j. Equation (4.36) holds that the total phase space volume of an infinitesimal *cloud packet* does not change over time; we can follow the trajectories of a small ensemble of particles as a function of time and will find that they always occupy the same amount of phase space. The corresponding density therefore does not change over time, $\frac{d\rho}{dt} = 0$.

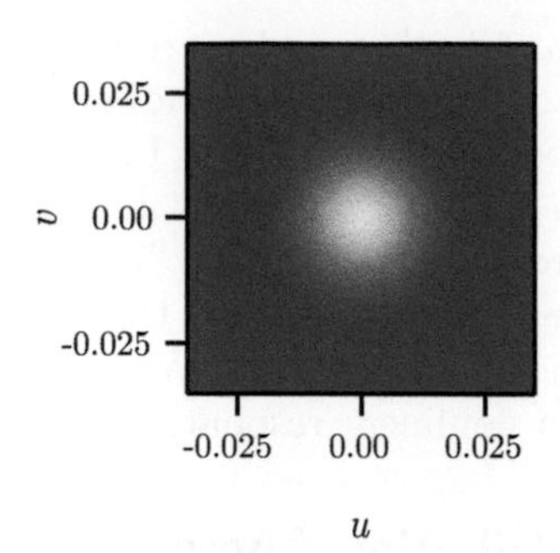

Fig. 4.7 Fokker–Planck calculation of the steady state of a damped harmonic oscillator with $\omega_0 = 100$, $Q = 100$, $m = 1$, $F_0 = 0$, and $\varsigma_D = 1$. The frame rotates at $\omega = \omega_0$. Bright yellow and dark blue correspond to high and low probability density, respectively.

We limit our notation here to the case of a one-dimensional potential, but the example could easily be extended to include y and z coordinates as well. Making use of eqns (1.2) and (1.3), Liouville's theorem can be rewritten as

$$\frac{\partial \rho}{\partial t} = -\sum_{j=1}^{N} \left(\frac{\partial \rho}{\partial x_j} \frac{\partial H}{\partial p_j} - \frac{\partial \rho}{\partial p_j} \frac{\partial H}{\partial x_j} \right) \equiv \{H, \rho\} \,, \tag{4.37}$$

where we have introduced the Poisson bracket $\{.,.\}$. Equation (4.37) describes the change of ρ at a particular position in phase space as a consequence of the potential landscape that each particle experiences.

4.3.2 The Fokker–Planck Equation

In order to apply a probability density approach to a single resonator under the influence of statistical forces, we modify eqn (4.37) as follows: (i) moving away from the picture of a cloud of many small particles, we interpret the density function as the probabilistic positions of a *single* particle in phase space. The finite probability is now a consequence of our limited knowledge of the actual system state caused by noise. (ii) Damping and diffusion forces are added to the EOM of the system to allow for fluctuations in phase space.

Under these new conditions, Liouville's theorem no longer holds. For instance, the presence of damping can cause all of the probability density to relax toward the origin of a potential well. Instead, we can make use of the **Fokker–Planck equation** [103, 104]

$$\frac{\partial \rho}{\partial t} = -\frac{\partial}{\partial x}(f_x \rho) - \frac{\partial}{\partial p}(f_p \rho) + \left(\frac{\partial}{\partial x^2} D_x + \frac{\partial}{\partial p^2} D_p \right) \rho \tag{4.38}$$

that includes two types of contributions: on the one hand, deterministic forces (*drift*) act in the directions of x and p with terms defined as $f_x = \dot{x}$ and $f_p = \dot{p}$ in the absence of fluctuating forces, $\varsigma_D^2 = 0$. On the other hand, force noise (*diffusion*) appears as terms $D_{x,p}$ that are proportional to the force noise PSD ς_D^2, cf. eqns (4.25) and (4.32).

The Fokker–Planck equation can be derived from the Îto process that we introduced in eqn (4.3), where the two types of forcing are equally present [101].

In the following, we will use the Fokker–Planck formalism for the visualization of noise-driven systems in the rotating frame. There, we define a rotating-phase-space density function $\rho(u,v,t)$ to replace $\rho(\mathbf{x},t)$ and obtain a corresponding Fokker–Planck equation

$$\frac{\partial \rho}{\partial t} = -\frac{\partial}{\partial u}(f_u \rho) - \frac{\partial}{\partial v}(f_v \rho) + \frac{\varsigma_D^2}{4\omega^2}\left(\frac{\partial}{\partial u^2} + \frac{\partial}{\partial v^2}\right)\rho. \tag{4.39}$$

For the case without diffusion and damping, the only forces remaining in our EOM are those exerted by the Hamiltonian, and eqn (4.39) becomes formally equivalent to a rotating-frame Liouville equation, cf. eqn (4.37). We then obtain $f_u = \dot{u} = \frac{\partial H_{\mathrm{rot}}}{\partial v}$ and $f_v = \dot{v} = -\frac{\partial H_{\mathrm{rot}}}{\partial u}$, yielding

$$\begin{aligned}\frac{\partial \rho}{\partial t} &= -\frac{\partial}{\partial u}(f_u \rho) - \frac{\partial}{\partial v}(f_v \rho) = -\frac{\partial}{\partial u}\left(\frac{\partial H_{\mathrm{rot}}}{\partial v}\rho\right) + \frac{\partial}{\partial v}\left(\frac{\partial H_{\mathrm{rot}}}{\partial u}\rho\right) \\ &= -\rho\frac{\partial}{\partial u}\frac{\partial H_{\mathrm{rot}}}{\partial v} - \frac{\partial H_{\mathrm{rot}}}{\partial v}\frac{\partial}{\partial u}\rho + \rho\frac{\partial}{\partial v}\frac{\partial H_{\mathrm{rot}}}{\partial u} + \frac{\partial H_{\mathrm{rot}}}{\partial u}\frac{\partial}{\partial v}\rho \\ &= -\frac{\partial H_{\mathrm{rot}}}{\partial v}\frac{\partial}{\partial u}\rho + \frac{\partial H_{\mathrm{rot}}}{\partial u}\frac{\partial}{\partial v}\rho,\end{aligned} \tag{4.40}$$

which corresponds exactly to the terms we found inside the sum in eqn (4.37).

For our linear resonator with external driving, the terms $f_{u,v}$ can be found from eqns (4.28) and (4.29) for $\Xi_u = \Xi_v = 0$ to be

$$f_u = -\frac{u\Gamma}{2} + \frac{v\left(\omega^2 - \omega_0^2\right)}{2\omega} + \frac{F_0 \sin\theta}{2m\omega} \tag{4.41}$$

and

$$f_v = -\frac{v\Gamma}{2} - \frac{u\left(\omega^2 - \omega_0^2\right)}{2\omega} - \frac{F_0 \cos\theta}{2m\omega}. \tag{4.42}$$

Equation (4.39) is a partial differential equation that can be solved numerically to follow the evolution of $\rho(u,v,t)$ as a function of time. It can also be solved for $\frac{\partial \rho}{\partial t} = 0$ to determine the stationary probability distribution of a given system. When numerically evaluating a probability density $\rho(u,v,t)$, we commonly discretize phase space, that is, divide it into $n \times m$ finite-sized bins with a defined value $\rho(u_n, v_m, t)$ at every step in time. Between steps, eqn (4.39) is applied to every bin individually, where spatial differentials are calculated from the differences between neighboring bins. In this way, the change of ρ over time is obtained at every position.

Using the Fokker–Planck equation, we plot the steady state of a noise-driven harmonic resonator in Fig. 4.7. The Fokker–Planck formalism is very useful to quantify statistical properties of a system. For instance, we can easily calculate the total displacement fluctuation power $\langle X^2 \rangle$ by integrating over the product of the probability distribution and the squared amplitude,

$$\langle X^2 \rangle = \int_{-\infty}^{\infty} \int_{-\infty}^{\infty} dudv(u^2 + v^2)\rho(u, v, t). \tag{4.43}$$

The result obtained for the data shown in Fig. 4.7 is $\langle X^2 \rangle = 10^{-4}$, in agreement with eqn (4.32).

Chapter summary

- In Chapter 4, we study the effect of fluctuating forces on the damped harmonic oscillator.
- When a damped harmonic oscillator is in contact with an environment of finite temperature, it must have a mean energy given by the **equipartition theorem**, cf. eqn (4.1). This energy is imparted on the oscillator through fluctuating forces that are connected to its damping via the **fluctuation-dissipation theorem**, cf. eqns (4.25) and (4.32) for formulations in the nonrotating and rotating frame, respectively.
- To study the oscillator's response to fluctuating forces, we introduce the **autocovariance**, cf. eqn (4.7), and the **power spectral density** (PSD), cf. eqn (4.8). The two representations are connected through the **Wiener–Khinchin theorem**.
- The **Parseval theorem** provides a link between the PSD and the variance of the displacement, cf. eqn (4.11).
- The slow-flow equations of a harmonic oscillator in a phase space rotating at ω_0 become decoupled, cf. eqns (4.34) and (4.35). When driven by thermal fluctuations, each quadrature of the oscillator performs a random walk modified by a linear centripetal force that is provided by the damping force.
- We introduce the probability distribution approach as a stepping stone between classical fluctuating physics and quantum physics. We start from **Liouville's theorem** for a large cloud of identical particles, cf. eqn (4.36). Applying this idea to the behavior of a single particle whose coordinates are subject to fluctuations and uncertainty, we arrive at the **Fokker–Planck equation** for a phase-space density function $\rho(u, v, t)$, cf. eqn (4.39). This equation can be solved analytically or numerically and allows us to follow the time evolution of a system.

Exercises

Check questions:

(a) In the response of an underdamped harmonic oscillator to force noise in Fig. 4.2(b), why do we see no fluctuations on timescales of a few oscillation periods?

(b) Can you formulate one or several new methods to evaluate Q in the presence of force noise?

(c) Compare the PSD of a noise-driven oscillator to the response function from eqn (1.14). How can you interpret the white force noise relative to the external force in a frequency sweep? What can you say about the phase response of the noise-driven system compared to eqn (1.15)?

(d) Why is the power x^2 often a more useful measure of a fluctuating degree of freedom than x itself?

(e) In what case can two PSDs be added linearly as $S_{x+y} = S_x + S_y$? Compare to the case of adding two variances σ_x^2 and σ_y^2.

(f) When (and why) are u and v correlated in a rotating-frame representation of a harmonic oscillator?

Tasks:

4.1 Use the code **Python Example 4** to numerically evolve the Îto process in the nonrotating frame, cf. eqn (4.3). Here, we define a temperature and apply the corresponding thermal force noise PSD to a harmonic oscillator. Use $\omega_0 = 1$, $Q = 100$, $T = 300$, and the resulting **sigma_m** as default values (and zero for everything else). Pay attention to the fact that **sigma_m** is defined as ς_D/m in this process. Test the fluctuation-dissipation theorem in the long-time limit, cf. eqn (4.25). How long does **tf** have to be to make the result representative of the stationary solution?

4.2 Study the response of $x(t)$ for short and long timescales and plot it together with the corresponding PSD. Can you understand all of the features you see? For instance, how does the PSD peak width manifest in the time domain?

4.3 Open the code **Python Example 3** to study the noise-driven harmonic oscillator in a frame rotating at $\omega = \omega_0$. Use various values of ω_0, m, and ς_D and visualize the result. Test the fluctuation-dissipation theorem in the long-time limit, cf. eqn (4.32). Investigate the influence of the sampling time step for the result by running a simulation with several steps per oscillation period or with several periods per step. In the frequency domain picture, what does a shorter time step correspond to?

4.4 Add a resonant force F_0 and observe what the combination of the coherent drive and the force noise produces. How do you have to modify eqn (4.32) to capture the role of the fluctuations in this situation? Test your intuition with a numerical experiment.

5

Parametric Resonators with Force Noise

We discussed in Section 4.2.3 that for the harmonic oscillator, the noise-induced fluctuations δu and δv average out in the long-time limit and that $\langle u\rangle$ and $\langle v\rangle$ remain unchanged. This, however, is not necessarily true in the presence of a nonlinearity and a parametric drive. As we saw in Chapter 3, these elements give rise to bifurcations, multistability, and hysteresis. In this chapter, we study the modification of the nonlinear parametric oscillator due to force fluctuations.

5.1 Multistability and Quasi-Stable Solutions

We found from the analysis of parametric symmetry breaking in Section 3.1.4 that the addition of a weak external harmonic force can under certain conditions change the bifurcation topology of a parametric oscillator. In Chapter 4, we saw that the force noise terms $\Xi_{u,v}$ in eqns (5.1) and (5.2) are also external forces whose amplitudes and phases fluctuate in time. This raises the question under which conditions the force noise can modify the stationary behavior of the system.

Close to a bifurcation point, the effect of force noise may be strong enough to displace the stationary solutions, thus inducing *hopping* between different attractors, see Fig. 5.1(a). Such hopping events will wash out the position of the bifurcation point in a space spanned by u, v, and ω. The event rate becomes strongly suppressed with increasing quasi-potential *barriers*, see Fig. 5.1(b). We will first limit our discussion to small noise and a system sufficiently far from bifurcation points to ignore such hopping, and then briefly touch upon hopping between attractors in Section 5.3.2.

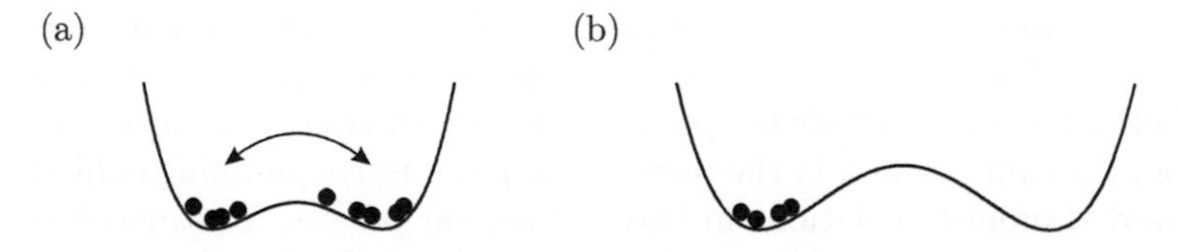

Fig. 5.1 Schematic representation of a bifurcated system with two attractors in the presence of force noise. (a) Close to a bifurcation point, several attractors are near to each other in phase space. Fluctuations induced by the force noise can cause stochastic hopping (at a finite escape rate) between the attractors. (b) For a large separation between the attractors, the escape rate between them is suppressed.

5.1.1 Slow-Flow and Fokker–Planck Equations

In Section 4.2.2, we saw how the slow-flow equations of a harmonic oscillator can be modified to account for the presence of force noise. Here, we generalize the equations to include a Duffing nonlinearity and a parametric drive, as we introduced in Chapters 2 and 3. The slow-flow equations we arrive at are [99, 100]

$$\begin{aligned}\dot{u} = &-\frac{u\Gamma}{2} - \frac{\left(u^3+uv^2\right)\eta}{8} - \frac{3\left(u^2v+v^3\right)\beta}{8\omega} + \frac{v\left(\omega^2-\omega_0^2\right)}{2\omega} \\ &+ \frac{\lambda\omega_0^2}{4\omega}\left(u\sin\psi - v\cos\psi\right) + \frac{F_0\sin\theta}{2m\omega} + \frac{\Xi_u}{m}, \end{aligned} \tag{5.1}$$

$$\begin{aligned}\dot{v} = &-\frac{v\Gamma}{2} - \frac{\left(v^3+u^2v\right)\eta}{8} + \frac{3\left(uv^2+u^3\right)\beta}{8\omega} - \frac{u\left(\omega^2-\omega_0^2\right)}{2\omega} \\ &- \frac{\lambda\omega_0^2}{4\omega}\left(u\cos\psi + v\sin\psi\right) - \frac{F_0\cos\theta}{2m\omega} + \frac{\Xi_v}{m}. \end{aligned} \tag{5.2}$$

With these modified slow-flow equations, the terms $f_{u,v}$ in the Fokker–Planck equation become

$$\begin{aligned}f_u = &-\frac{u\Gamma}{2} - \frac{\left(u^3+uv^2\right)\eta}{8} - \frac{3\left(u^2v+v^3\right)\beta}{8\omega} + \frac{v\left(\omega^2-\omega_0^2\right)}{2\omega} \\ &+ \frac{\lambda\omega_0^2}{4\omega}\left(u\sin\psi - v\cos\psi\right) + \frac{F_0\sin\theta}{2m\omega}, \end{aligned} \tag{5.3}$$

and

$$\begin{aligned}f_v = &-\frac{v\Gamma}{2} - \frac{\left(v^3+u^2v\right)\eta}{8} + \frac{3\left(uv^2+u^3\right)\beta}{8\omega} - \frac{u\left(\omega^2-\omega_0^2\right)}{2\omega} \\ &- \frac{\lambda\omega_0^2}{4\omega}\left(u\cos\psi + v\sin\psi\right) - \frac{F_0\cos\theta}{2m\omega}. \end{aligned} \tag{5.4}$$

In the following, we analyze the interplay between external driving, parametric pumping, and noise in the regimes below and above the parametric driving threshold.

5.2 Parametric Amplification below Threshold

We begin by looking at the effect of relatively weak parametric pumping and $F_0 = 0$, where nonlinearities are typically negligible. In Fig. 5.2, we compare the rotating-frame behavior of a resonator in the long-time limit without and with parametric pumping below threshold. While the purely noise-driven resonator has a random phase distribution, the same system in the presence of parametric pumping exhibits reduced and increased standard deviations in two orthogonal phases. The modified standard deviations are a consequence of the phase-dependent effective damping coefficients caused by the pump. As seen in Fig. 5.2(b), the parametrically pumped fluctuations become **correlated** in phase, that is, they prefer one particular phase axis.

In quantum mechanics, this phenomenon is called **squeezing** when it leads to a state that has an uncertainty smaller than that of the quantum ground state in one quadrature. When applied to a thermal state (where the uncertainly in both

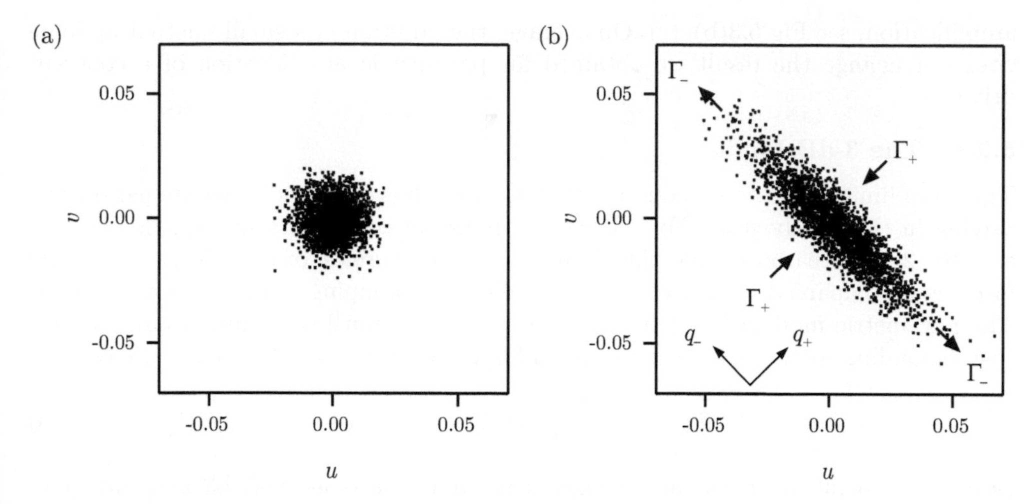

Fig. 5.2 (a) Numerical simulation of a resonator driven by force noise with $\omega_0 = 100$, $Q = 100$, $m = 1$, $F_0 = 0$, $\varsigma_D = 1$, $\beta = 0$, $\eta = 0$, and $\lambda = 0$ in a frame rotating at $\omega = \omega_0$. This is the same data as in Fig. 4.6. (b) The same simulation with a parametric pump modulation depth $\lambda = 0.9\lambda_{\rm th}$ and phase $\psi = 0$. Black arrows indicate the phases with increased (Γ_+) and decreased (Γ_-) damping, cf. eqns (5.6) and (5.5), respectively.

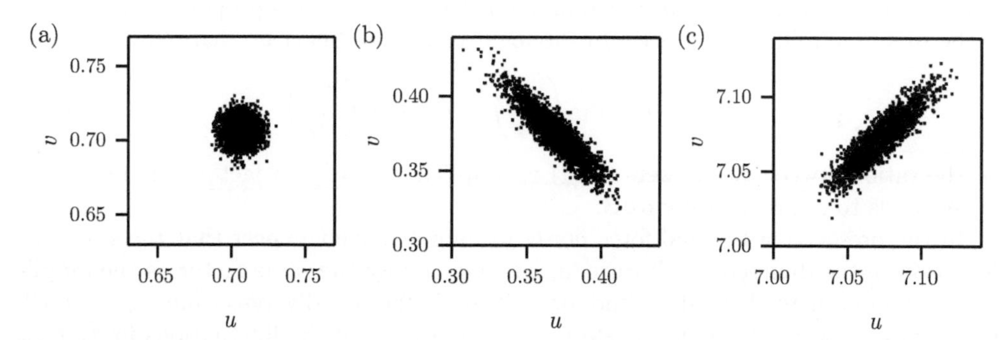

Fig. 5.3 (a) Numerical simulation of a resonator driven by coherent and fluctuating forces with $\omega_0 = 100$, $Q = 100$, $m = 1$, $F_0 = 100$, $\theta = \frac{3}{4}\pi$, $\varsigma_D = 1$, $\beta = 0$, $\eta = 0$, and $\lambda = 0$. The evolution was simulated over a time of 5000. (b) The same simulation with a parametric pump modulation depth $\lambda = 0.9\lambda_{\rm th}$ and phase $\psi = 0$ or (c) with $\psi = \pi$. Notice that the coherent displacement in phase space is (de-)amplified along with the displacement noise. All frames rotate at $\omega = \omega_0$.

quadratures is much larger than that of the ground state), some communities refer to the phenomenon as **classical squeezing** [22]. In the following, we use the notation q_- and q_+ to denote the amplified and damped quadratures, as indicated in Fig. 5.2(b).

As we saw in Chapter 3, parametric amplification can also be applied to a resonator that is subject to a coherent, external drive. In Fig. 5.3(a), we show the response of a resonator to a combination of a coherent drive and force noise. In the presence of a parametric pump, both driving sources are subject to phase-dependent parametric amplification, see Fig 5.3(b)–(c). On average, the addition of a small fluctuating force does not change the result we obtained for parametric amplification of a coherent drive.

5.2.1 The 3-dB Limit

There is a limit to the noise reduction that can be achieved with degenerate parametric driving in the steady state. For a driven response in the quadrature q_+, this limit is a factor 2. The reason behind this limit is that the increase of the effective damping in q_+ is accompanied by a decrease of the effective damping in q_-, cf. Section 3.1.1. The parametric modulation depth λ can be increased until the reduced damping Γ_- (corresponding to q_-) reaches zero, which happens for $\lambda_{\text{th}} = 2/Q$ because there

$$\Gamma_- = \Gamma - \frac{\lambda_{\text{th}}\omega_0}{2} = \Gamma - \frac{2\omega_0}{2Q} = 0\,. \tag{5.5}$$

Beyond this point, the harmonic parametric resonator becomes unstable and undergoes a transition to a high-amplitude nonlinear phase state. For a phase rotated by 90 degrees, the damping in the quadrature q_+ can be increased up to

$$\Gamma_+ = \Gamma + \frac{\lambda_{\text{th}}\omega_0}{2} = \Gamma + \frac{2\omega_0}{2Q} = 2\Gamma\,. \tag{5.6}$$

The increase of the damping by a maximum factor of 2 leads to a proportional decrease of the driven amplitude response. In the logarithmic decibel scale that is defined as

$$R_{dB} = 10\log_{10}\left(\frac{P_1}{P_2}\right) = 20\log_{10}\left(\frac{A_1}{A_2}\right) \tag{5.7}$$

for the ratio between two powers ($P_{1,2}$) or amplitudes ($A_{1,2}$), a factor 2 in amplitude corresponds to approximately 6 dB.

In the presence of thermal force noise, we might naively expect that the standard deviation of the damped quadrature, σ_+, is reduced by the same factor as the amplitude response. Instead, what we find for $\lambda \to \lambda_{\text{th}}$ is that σ_- diverges while $\sigma_+^2 \to \sigma_0^2/2$, where $\sigma_0 = \sigma_{u,v}(\lambda = 0)$. It is therefore the variance σ_+^2 of the fluctuations in q_+ that is reduced by a factor 2, not their standard deviation σ_+ [105]. On a logarithmic scale, a factor of 2 in variance (or power) is roughly 3 dB.

Why is the driven response damped by up to 6 dB, while fluctuations in the same quadrature can only be decreased by 3 dB? To understand this, we take a look at the quadrature response functions. Following Ref. [105], we write down the thermal displacement noise PSDs of the fluctuations in $q_\pm$ as

$$S_\pm(\omega/2\pi) = \frac{S_\pm(\omega)}{2\pi} = \frac{k_BT\Gamma}{m\omega_0^2}\frac{1}{\omega^2 + \Gamma_\pm^2/4}\,, \tag{5.8}$$

in analogy to the thermal displacement PSD of x in eqns (4.9) and (4.10). The variances $\sigma_\pm^2$ are calculated from the integration of $S_\pm$ and yield

$$\sigma_\pm^2 = \int_{-\infty}^{\infty} S_\pm(\omega)d\omega = \frac{k_BT}{m\omega_0^2}\frac{\Gamma}{\Gamma_\pm}\,. \tag{5.9}$$

(a)

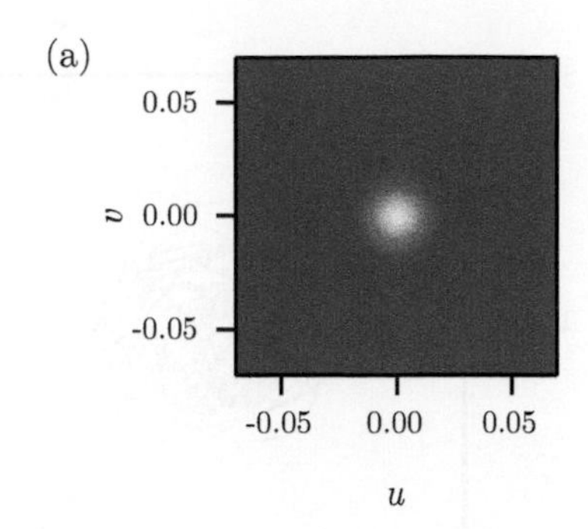

(b)

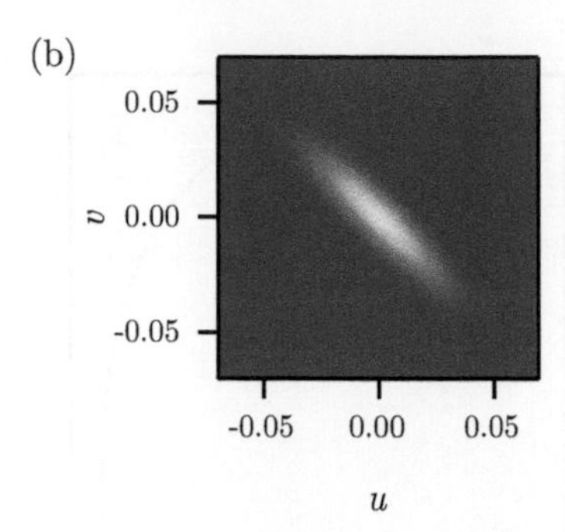

Fig. 5.4 (a) Fokker–Planck calculation of the stationary distribution of a resonator with $\omega_0 = 100$, $Q = 100$, $m = 1$, $F_0 = 0$, $\varsigma_D = 1$, $\beta = 0$, $\eta = 0$, and $\lambda = 0$, cf. eqn (4.39) with eqns (5.3) and (5.4). The frame rotates at $\omega = \omega_0$. Bright yellow and dark blue correspond to high and low probability density, respectively. From eqn (4.43) we obtain $\sigma_X^2 = 10^{-4}$. (b) The same calculation with $\lambda = 0.9\lambda_{\text{th}}$ and $\psi = 0$ yields $\sigma_X^2 = 5.3 \times 10^{-4}$.

For $\lambda = 0$, eqn (5.9) reduces to the rotating-frame analogue to the equipartition theorem in eqn (4.33) after replacing u and v with the rotated quadratures q_+ and q_-. For $\lambda = \lambda_{\text{th}}$, the fluctuation variance in q_+ is reduced by a factor 2 and the variance in q_- diverges. There are two competing effects at work here; changing the effective damping coefficient with the parametric drive damps the peak power response by a maximum factor of 4. At the same time, though, the full-width-at-half-maximum of the peak increases by a factor of 2. The integral over the entire spectrum is therefore decreased by a total factor of 2. The parametric drive breaks the equipartition theorem in eqn (4.9) by changing the effective Γ in the susceptibility function χ without modifying the thermal force noise in eqn (4.25) accordingly. As a consequence of the increased peak width, the total (integrated) noise standard deviation σ_+ can only be reduced by a factor $\sqrt{2}$, that is, 3 dB, while the resonant response that is probed with a coherent drive F_0 can be suppressed by up to a factor 2, that is, 6 dB.

Other methods for classical squeezing, for instance pulsed squeezing or combinations of squeezing and feedback damping, do not necessarily have a 3 dB limit [105]. Furthermore, the use of narrow-band filters can make a measurement sensitive to only the central part of the response curve. In this case, the bandwidth increase we discussed above is irrelevant and the maximum reduction of the recorded (filtered) variance is 6 dB [22].

We can also use eqn (5.9) to predict that parametric amplification applied to a thermal state increases the total fluctuating displacement power $\langle X^2 \rangle$. The sum $\langle X^2 \rangle = \langle q_-^2 \rangle + \langle q_+^2 \rangle = \sigma_-^2 + \sigma_+^2$ is given by

$$\langle X^2 \rangle = \frac{k_B T}{m\omega_0^2} \frac{\Gamma}{\Gamma_- \Gamma_+} (\Gamma_- + \Gamma_+) \,. \tag{5.10}$$

A numerical example of this effect is presented in Fig. 5.4. Using eqn (4.43) and eqn (5.10), we consistently obtain an increase of $\langle X^2 \rangle$ by roughly a factor 5 between cases with $\lambda = 0$ and those with $\lambda = 0.9\lambda_{\text{th}}$.

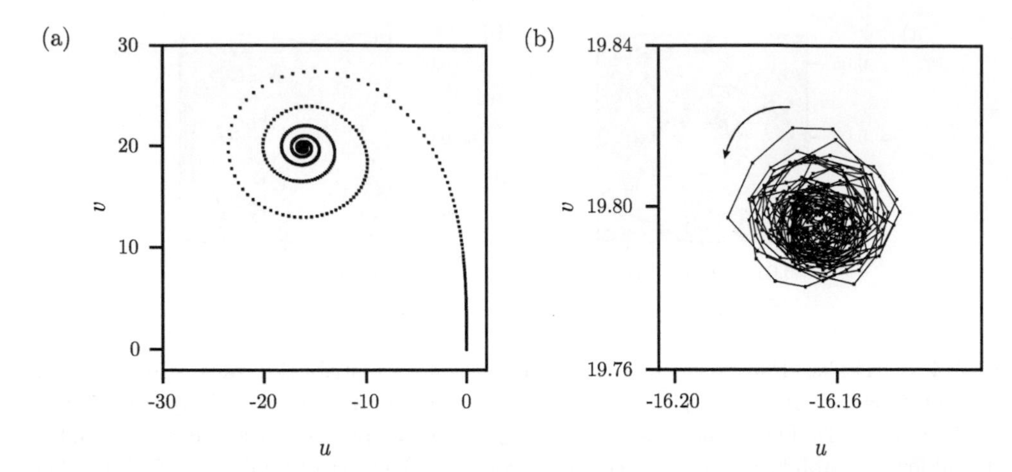

Fig. 5.5 (a) Numerical simulation of a resonator ringing up to a phase state with $\omega_0 = 100$, $Q = 100$, $m = 1$, $F_0 = 0$, $\varsigma_D = 1$, $\beta = 1$, $\eta = 0$, $\lambda = 5\lambda_{\text{th}}$, and $\psi = -\pi/2$, cf. eqns (5.1) and (5.2). The evolution was simulated over a time of 5000 and the frame rotates at $\omega = \omega_0$. Note that the initial condition was $u = v = 0$ and that the random fluctuations caused by the force noise determined which phase state is chosen. (b) Fluctuations of a resonator around the attractor. Parameters are the same as in (a) with a time step interval of 0.1. Lines connecting the data points serve to visualize the trajectories, and an arrow indicates the direction of rotation.

5.3 Parametric Pumping Above Threshold

A fluctuating force also has important consequences for a system that is parametrically pumped above threshold ($\lambda > \lambda_{\text{th}}$). In the absence of a coherent driving force, the fluctuating displacement in response to a force noise is sufficient to provide a nonzero starting condition that allows the resonator to ring up into a phase state, see Fig. 5.5(a). The force noise is therefore the catalyst for the **spontaneous time-translation symmetry breaking** of the resonator, by which we mean that the system chooses one of the phase states randomly. Note that for classical force noise, the randomness is merely a consequence of the fluctuating displacement in $u(t)$ and $v(t)$.

In Fig. 5.5(a), we observe that the resonator spirals around the attractor as it converges toward it, as we have already seen in Fig. 3.3. The force noise has little impact as long as u and v are far enough from the attractor, that is, the parametric pump dominates the overall dynamics far from the stationary points. After the ringup transients have died down, however, the constant terms in our slow-flow equations balance each other. At this point, only the force noise will cause the resonator state to fluctuate around the attractor, see Fig. 5.5(b). For weak noise, we can understand this behavior in terms of a parametric symmetry breaking force that effectively shifts the coordinates of the equilibrium point, cf. Fig. 3.6(b). Here, however, the force is stochastic, such that its amplitude and phase change on a timescale of τ_0, cf. Chapter 4. The random force kicks cause the resonator to undergo displacement fluctuations δu and δv that are typically much smaller than the amplitude X of the phase state.

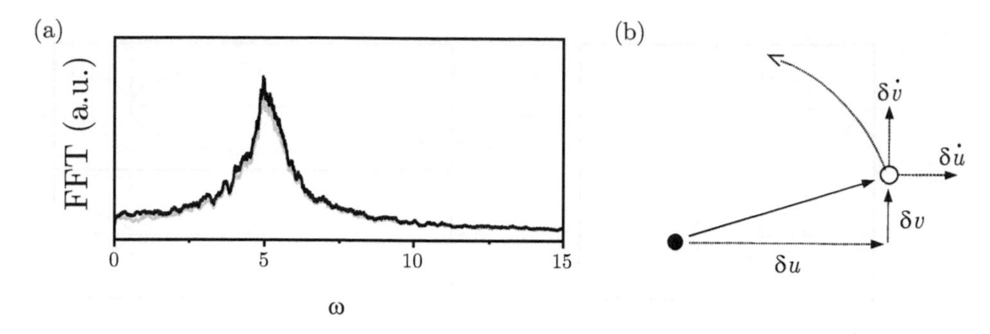

Fig. 5.6 (a) Fast Fourier transform (in arbitrary units) of δu (gray) and δv (black) of the data shown in Fig. 5.5(b). Slow oscillations in both quadratures lead to peaks in the spectrum. (b) Schematic illustration of resonator fluctuations. The black dot marks the attractor with coordinates u_{eq} and v_{eq}. A fluctuating displacement δu leads to relaxation toward the attractor as well as oscillation around the attractor due to the coupled degrees of freedom.

At a first glance, this situation appears analogous to the one of the thermal harmonic oscillator in Section 4.2.3, with the difference that the attractor here is at a high amplitude. Upon closer inspection, though, we notice some crucial differences. Due to the nonlinearities β and η in eqns (5.1) and (5.2), the averaged degrees of freedom u and v are now coupled. If we pump the resonator at $\omega = \omega_0$ into a phase state, the timescales of thermal fluctuations δu and δv will depend not only on Γ but also on some rotating-frame oscillation frequency given by the nonlinear terms [106]. In Fig. 5.6(a), we track these slow rotations through the fast Fourier transforms of δu and δv; it is important to stress that these oscillations are not governed by ω_0 but directly by the nonlinear nature of the quasi-potential H_{rot}, cf. eqn (3.22).

5.3.1 The Method of Characteristic Exponents

In order to gain a better picture of the rotating-frame fluctuations, let us consider a parametric oscillator phase state obtained for $F_0 = \eta = 0$, and let us denote the coordinates of its equilibrium point as u_{eq} and v_{eq}. Thermal force noise displaces the system slightly from the center of the attractor, see Fig. 5.5(b). What forces will the system experience? Ignoring the weak force noise for now, we realize that the differential equation to describe the short-time dynamics is the same as what we used for our stability analysis in Section 3.1.3. Namely, we approximate the motion with a linear equation for the fluctuations of the form [107]

$$\begin{bmatrix} \delta\dot{u} \\ \delta\dot{v} \end{bmatrix} = M \begin{bmatrix} \delta u \\ \delta v \end{bmatrix}, \tag{5.11}$$

where the Jacobian matrix has the entries (for $\eta = 0$)

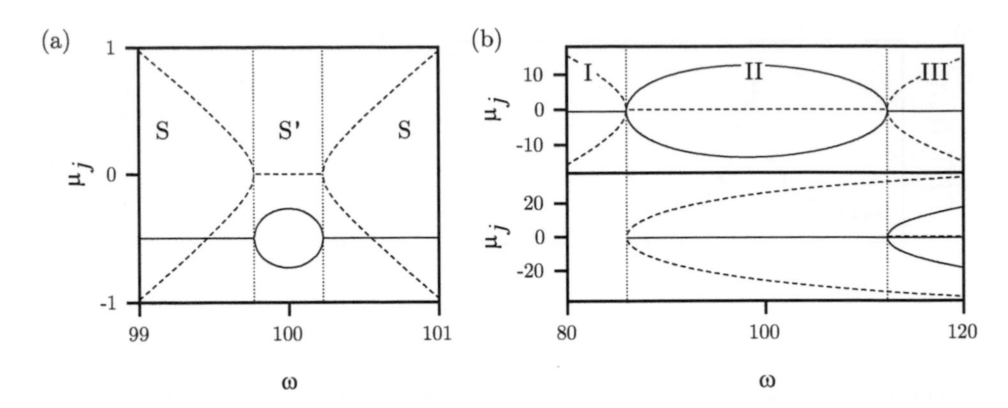

Fig. 5.7 Characteristic exponents of fluctuations of a parametrically driven resonator around different attractors with $\omega_0 = 100$, $Q = 100$, $m = 1$, $F_0 = 0$, $\beta = 1$, and $\eta = 0$. Real and imaginary parts are drawn as solid and dashed lines, respectively. (a) Frequency dependence of μ_j below threshold, $\lambda = 0.9\lambda_{\text{th}}$, where $u_{\text{eq}} = v_{\text{eq}} = 0$ is the sole attractor. The real part equals $-\Gamma/2 = -0.5$ outside the region S' (cf. Fig. 3.5). Within S', parametric amplification leads to increased and decreased effective damping $\Gamma_{+/-}$. Far from ω_0, the imaginary parts of μ_j reflect the detuning $\omega - \omega_0$ that leads to an apparent oscillation in the rotating frame. Inside S', the resonator becomes locked to the parametric drive. (b) Above threshold with $\lambda = 5\lambda_{\text{th}}$, there are 1, 2, and 3 stable solutions in the regions I, II, and III, respectively (cf. Fig. 3.2). The upper panel shows the zero-amplitude solution. This solution strongly resembles the one in (a), with the notable exception that it has one real value > 0 in region II and thus becomes unstable inside the Arnold tongue. The lower panel shows two additional solutions that appear in regions II and III. The stable solutions with a real part of $-\Gamma/2$ and frequency-dependent imaginary parts are the phase states, cf. Fig. 5.5(b) and 5.6(a). In region III, we find additional unstable states that merge with the phase states at the boundary to region IV, cf. Fig. 3.2.

$$
\begin{aligned}
M_{11} &= \frac{\partial f_1}{\partial \delta u} = -\frac{6uv\beta + 4\Gamma\omega - 2\lambda\omega_0^2 \sin\psi}{8\omega}, \\
M_{12} &= \frac{\partial f_1}{\partial \delta v} = -\frac{3\beta(u^2 + 3v^2) + 4(\omega_0^2 - \omega^2) + 2\lambda\omega_0^2 \cos\psi}{8\omega}, \\
M_{21} &= \frac{\partial f_2}{\partial \delta u} = \frac{3\beta(3u^2 + v^2) + 4(\omega_0^2 - \omega^2) - 2\lambda\omega_0^2 \cos\psi}{8\omega}, \\
M_{22} &= \frac{\partial f_2}{\partial \delta v} = -\frac{-6uv\beta + 4\Gamma\omega + 2\lambda\omega_0^2 \sin\psi}{8\omega},
\end{aligned}
\tag{5.12}
$$

with $u \approx u_{\text{eq}}$ and $v \approx v_{\text{eq}}$.

We can quantify the contributions corresponding to damping and oscillation around an attractor with the method of characteristic exponents μ_j, which are the eigenvalues of the Jacobian matrix. We employed this method in Section 3.1.3 to determine whether an equilibrium point is stable, see eqn (3.21). There, we found as the criterion for stability that the real parts of the characteristic exponents μ_j at an equilibrium

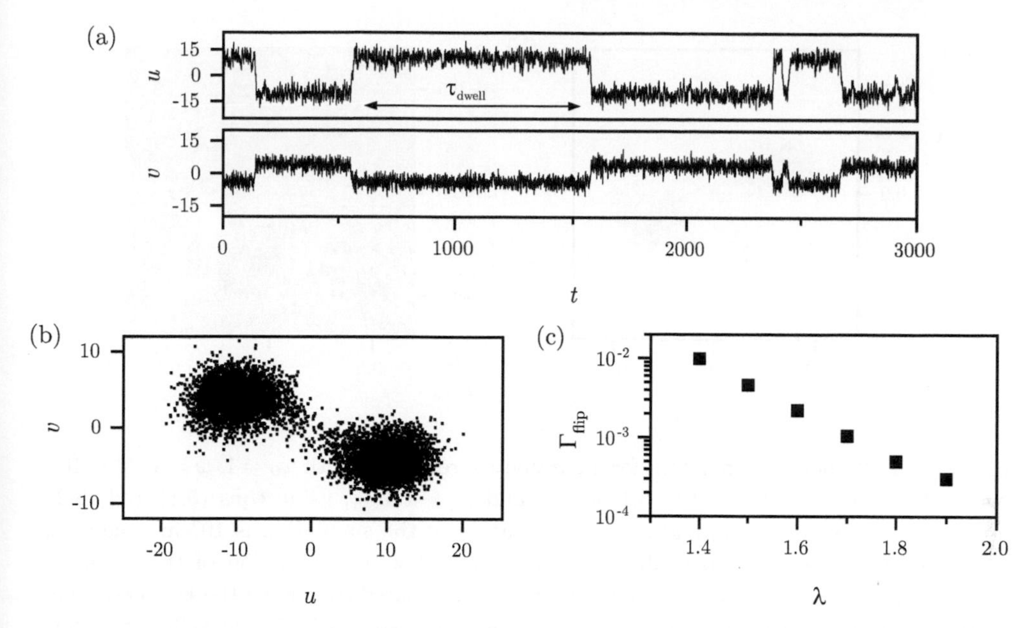

Fig. 5.8 Numerical simulation of phase state hopping due to large force noise. (a) Plots of u and v as a function of time t, for parameters $\omega_0 = 100$, $Q = 100$, $m = 1$, $F_0 = 0$, $\varsigma_D = 300$, $\beta = 1$, $\eta = 0$, $\lambda = 1.5\lambda_{th}$, and $\psi = 0$. The evolution is simulated over a time of 5000 and the frame rotates at $\omega = \omega_0$. (b) Phase state portrait of the parametric oscillator over the entire simulation time. (c) Averaged hopping rate between the two phase state attractors as a function of λ. Each rate was estimated from simulated evolutions over a time of 20000 with a frame rotating at $\omega = \omega_0$.

point must all be negative. Here, we keep both the real and imaginary parts of the exponents, as we did in the nonrotating frame in eqn (1.11). The real (imaginary) part of a characteristic exponent μ characterizes the motion toward (around) the center of an attractor and determines the width (position) of the peak appearing in the Fourier transform of the fluctuations, see Fig. 5.6(a).

For a given equilibrium point that fulfills $\dot{u} = \dot{v} = 0$, we use eqn (3.21) to find the characteristic exponents μ_j. Plotting the real and imaginary parts of those exponents as a function of frequency (for otherwise fixed parameters) yields plots as shown in Fig. 5.7.

5.3.2 Hopping Between Attractors

Strong force noise can lead to random hopping between the attractors of different stable states, see Fig. 5.8 [108]. The resonator dwells in one particular state for a time τ_{dwell} before hopping to a different state. Each hopping event is the consequence of a random walk in phase space that, in some cases, gradually carries the system beyond a barrier in the quasi-potential topography between two attractors. The times at which noise-induced hopping events occur cannot be predicted deterministically; only their statistical probability or average rate $\Gamma_{flip} = \tau_{dwell}^{-1}$ can be calculated. The dwell time

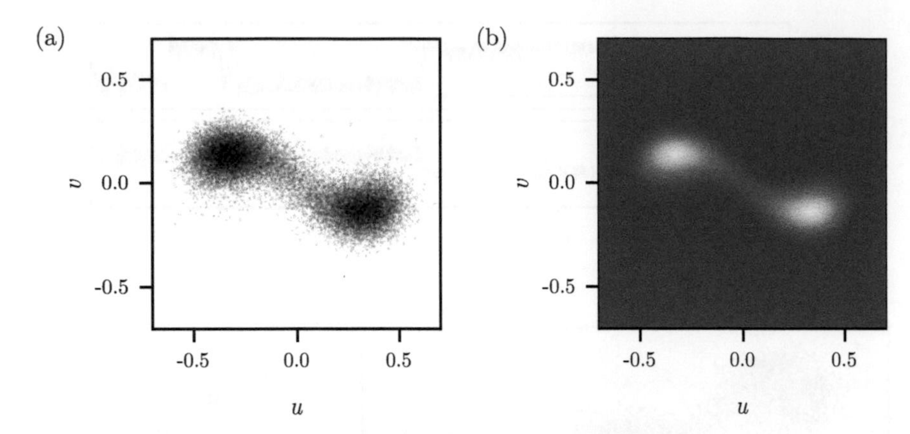

Fig. 5.9 (a) Numerical simulation for a parametric resonator with $\omega_0 = 1$, $\omega = 1$, $Q = 100$, $m = 1$, $F_0 = 0$, $\varsigma_D = 0.01$, $\beta = 0.1$, $\eta = 0$ and $\lambda = 1.5\lambda_{\text{th}}$, using eqns (5.1) and (5.2). The frame rotates at $\omega = \omega_0$ and the total duration of the simulation is 10^6 in time steps of 40. The data points are partially transparent to allow an estimation of their density. (b) Stationary solution $\rho(u, v, t)$ of the same system calculated by solving the Fokker–Planck equation, eqn (4.39), with the coefficients from eqns (5.3) and (5.4). Bright yellow and dark blue correspond to high and low probability density, respectively.

reflects the effective height of the barrier. Since the probability of a process depends on the quasi-potential barrier that the system has to overcome, one can use the notion of an activation energy to estimate Γ_{flip} under certain conditions [49].

Examples of the hopping rate obtained from numerical simulations are shown in Fig. 5.8(c). Typically, the rate depends in an approximately exponential fashion on the ratio between the attractor separation and the noise power, and therefore on the modulation depth λ, the noise PSD ς_D, and the detuning $\omega - \omega_0$. In response to the quasi-potential shape, the trajectories predominantly align along *phase space channels* that offer a minimal effective barrier [109]. Measuring the statistical density of trajectories can therefore serve as a probe of the quasi-potential landscape itself.

5.3.3 Multistability in the Probability Density Picture

Noise-activated hopping has important effects on the way we perceive systems with several attractors. In Fig. 5.9, we provide a side-by-side comparison between the stationary distribution of a parametric oscillator calculated with the stochastic differential equations for a sample path, and with the Fokker–Planck equation. The former is calculated using eqns (5.1) and (5.2) in discrete steps as a function of time. The latter is based on a numerical solution of eqn (4.39) after inserting eqns (5.3) and (5.4). The comparison shows the same statistical probability of finding the system at a particular position in phase space. However, the probability density in Fig. 5.9(b) clearly retains no information on the time scale of hopping, or on the random paths during each hop. In the probability density $\rho(u, v, t)$, we can draw no clear distinction between the various stable states of the system.

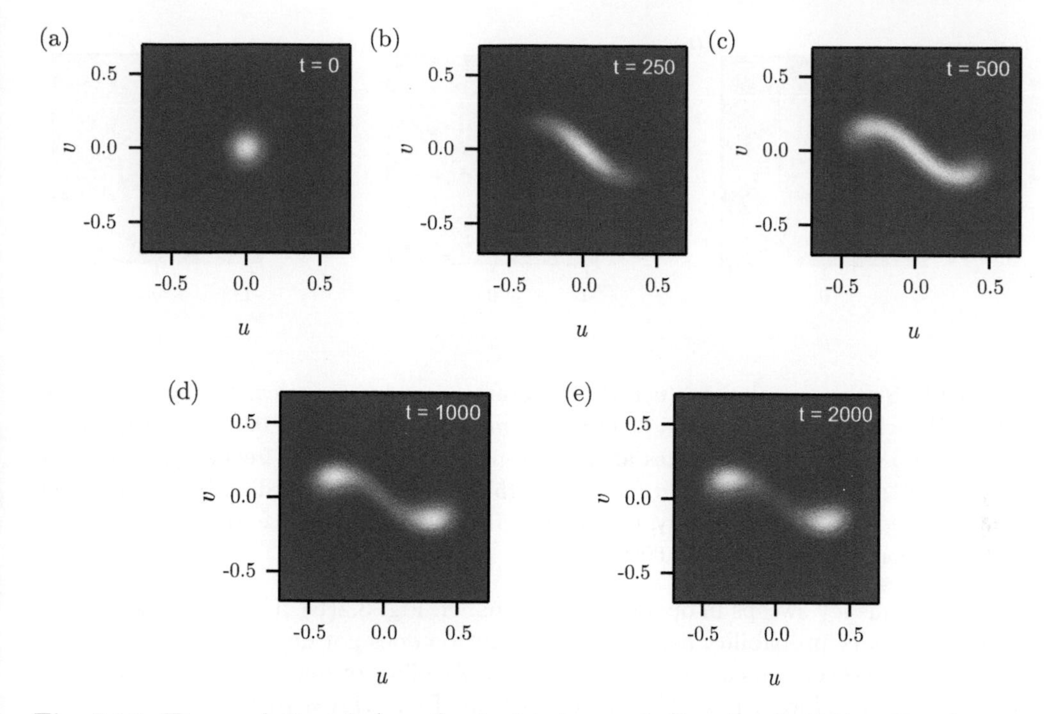

Fig. 5.10 Time evolution of $\rho(u,v,t)$ calculated numerically using the Fokker–Planck equation, cf. eqn (4.39), for a parametric resonator resonator with $\omega_0 = 1$, $\omega = 1$, $Q = 100$, $m = 1$, $F_0 = 0$, $\varsigma_D = 0.01$, $\beta = 0.1$, $\eta = 0$, and $\lambda = 1.5\lambda_{\rm th}$. The system is initialized in a thermal state and the parametric pump is switched on at $t = 0$. The frame rotates at $\omega = \omega_0$. Bright yellow and dark blue correspond to high and low probability density, respectively. The panels show the state at (a) $t = 0$, (b) $t = 250$, (c) $t = 500$, (d) $t = 1000$, and (e) $t = 2000$.

The difference between a sample-path simulation and the probability density approach becomes more acute when solving the Fokker–Planck equation as a function of time. In Fig. 5.10, we initialize a nonlinear resonator in its thermal steady state (for $\lambda = 0$) and then follow its evolution under the influence of a strong parametric pump ($\lambda > \lambda_{\rm th}$). We find that the state of the resonator is symmetric around the origin at any time. The probability density approach provides us with a solution that is composed out of all possible trajectories of a system. The notion of spontaneous symmetry breaking at bifurcation points is therefore lost. This is in stark contrast to the deterministic behavior we studied in Chapter 3, and in particular in Fig. 3.3: there, the symmetry was broken for each trajectory as the system rang up to one of the phase states.

By exploring all possible solutions, the probability density picture affords us an alternative view on bifurcation points. Instead of the qualitative information whether a state is stable or not, we can now gain a quantitative understanding or how stable a solution is. In Fig. 5.11, we see three stationary solutions calculated in region III, that is, outside of the Arnold tongue where the phase states coexist with an attractor at zero amplitude. In a deterministic system with $\varsigma_D = 0$, this coexistence leads to a hysteresis

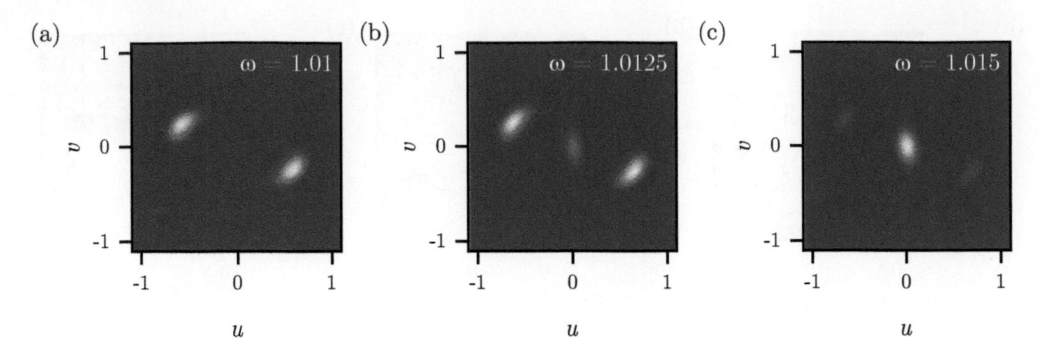

Fig. 5.11 Stationary solution $\rho(u, v, t)$ of a parametric resonator calculated by solving the Fokker–Planck equation with $\omega_0 = 1$, $Q = 100$, $m = 1$, $F_0 = 0$, $\varsigma_D = 0.01$, $\beta = 0.1$, $\eta = 0$, and $\lambda = 1.5\lambda_{\text{th}}$. The frame rotates at half the parametric pumping frequency, that is, at (a) $\omega = 1.01$, (b) $\omega = 1.0125$, and (c) $\omega = 1.015$. Bright yellow and dark blue correspond to high and low probability density, respectively. In the absence of noise, the zero-amplitude solution becomes stable for $\omega \geq 1.0056$.

between frequency sweeps in opposite directions, cf. Fig. 3.2(b). This hysteresis is lost in the stationary probability density, as the solution does not depend on any previous trajectory. However, we can observe how the probability of finding the system in the zero-amplitude solution gradually increases from Fig. 5.11(a) to (c). This indicates that the solution, although stable in the absence of noise, offers only a small barrier against escape close to the bifurcation point. For larger detuning, a stronger barrier develops, corresponding to a more pronounced maximum in the quasi-Hamiltonian illustration in Fig. 3.4. At the same time, the probability of finding the system in the phase states decreases, even though the solutions remain technically stable for $\omega \to \infty$ and $\eta = 0$.

5.4 Hierarchy of Relevant Timescales

At this point, we are well positioned to gain an overview of different timescales in our system. In our treatment of resonators, we explicitly or implicitly implied the hierarchy

$$T_p \ll \tau_0 \ll \tau_{\text{dwell}}\,, \tag{5.13}$$

where $T_p = \nu_0^{-1}$ is the natural (unforced) period of the oscillator, $\tau_0 = 2/\Gamma$ is the decay time, and τ_{dwell} is the average dwell time of the system in one particular attractor. The ordering in eqn (5.13) is fulfilled in a system with $Q \gg 1$ (underdamped case) and small enough noise.

The phenomena observed in simulations or experiments depend on the sampling time step t_{samp}. For $t_{\text{samp}} \ll T_p$, we can follow the oscillations of an underdamped resonator in the nonrotating frame, cf. Fig. 1.4. This type of sampling offers the largest amount of information at the expense of large data sets and costly calculations for relatively short observation times. The averaging method offers a much more efficient method for following the sample path over long times. Here, the time steps are typically chosen to be $T_p < t_{\text{samp}} < \tau_0$. We therefore neglect dynamics within a period but observe transients and slow fluctuations, cf. Figs. 3.3 and 5.5 for examples.

The dwell time τ_{dwell} was introduced in Section 5.3.2 to account for the presence of noise, which can induce hopping between multiple attractors. In the deterministic models studied in Chapters 1 to 3, no noise was present, resulting in $\tau_{\text{dwell}} \to \infty$. In this case, the system follows a particular attractor for arbitrarily long times as long as it remains stable. In other words, the full duration of a measurement (or simulation) needs to be much shorter than the dwell time ($t_S \ll \tau_{\text{dwell}}$) in order to observe phenomena such as hysteresis and spontaneous symmetry breaking.

If the noise becomes strong enough, there will be hopping during a measurement, $t_S > \tau_{\text{dwell}}$. This leads to situations as shown in Fig. 5.8 where various attractors are explored by the system. As a consequence, the symmetry between the phase states is restored and no hysteresis can be identified in a frequency sweep. For $t_S \gg \tau_{\text{dwell}}$, we arrive at probabilistic representations where all dynamics on timescales of T_p or τ_0 are ignored. The steady-state solution of the Fokker–Planck equation is the most extreme example of this approach, corresponding to an experimental situation $t_S \to \infty$.

Chapter summary

- In Chapter 5, we apply fluctuating forces to the nonlinear, parametric oscillator.
- We start by formulating the slow-flow equations for the parametric oscillator with force noise, cf. eqns (5.1) and (5.2), and the corresponding Fokker–Planck terms, cf. eqns (5.3) and (5.4). In the absence of an external drive or parametric pump, the effect of the nonlinearities is usually negligible and the system can be approximated by the thermally driven harmonic oscillator from Chapter 4.
- For a parametric pump with $\lambda < \lambda_{\text{th}}$, the standard deviation of the fluctuating displacement increases and decreases in two orthogonal phases, cf. Fig. 5.2. This phenomenon is a consequence of the parametric amplification discussed in Chapter 3 and is referred to as classical **squeezing**. The maximum reduction of the displacement is between 3 dB and 6 dB, depending on the bandwidth of a measurement, cf. the discussion around eqn (5.9). In general, squeezing increases the total variance of the fluctuations σ_X^2, cf eqn (5.10).
- For $\lambda > \lambda_{\text{th}}$, the oscillator rings up to the phase states discussed in Chapter 3. Here, the role of the force noise is threefold: first, when starting from the initial conditions $x = \dot{x} = 0$, force noise causes **spontaneous symmetry breaking** via random displacement fluctuations. Second, after a ringup to a stable attractor, the noise causes fluctuations around the attractor that can be studied with the method of **characteristic exponents**, cf. Fig. 5.7 and eqn (5.11). Third, strong noise can lead to **hopping** between the attractors on a timescale τ_{dwell}, cf. Fig. 5.8. The hopping amplitude depends strongly on the phase state amplitudes and on the noise PSD.
- On timescales much longer than the typical dwell time τ_{dwell}, hopping between solutions causes the **disappearance of hysteresis**, as all possible solutions are sampled with a certain probability. In this situation, the probability density approach based on the Fokker–Planck formalism offers a useful way to study the system, cf. Figs 5.10 to 5.11.

Exercises

Check questions:

(a) Looking at eqn (3.22), what is the shape of the quasi-potential of a harmonic oscillator as a function of detuning and damping? How does it change with a parametric pump below threshold?

(b) For a parametric oscillator ($\lambda > \lambda_{\rm th}$), force noise can activate hopping between the stable solutions, cf. Section 5.3.2. In what aspects is the parametric oscillator more complex than a static double well potential?

(c) In the rotating frame, what is the role of the measurement bandwidth (i.e. the range of frequencies away from ω_0 that is observed) on the maximum achievable noise reduction with parametric squeezing?

(d) Why do the fluctuations around a phase state have a characteristic frequency? What role does damping play for the corresponding fluctuation spectrum?

(e) When is it useful to use the Fokker–Planck steady-state probability density $\rho(u,v)$ instead of a numerical simulation of $x(t)$ to analyze a system? When do you prefer the latter? When can you combine the best of both methods with the time-dependent Fokker–Planck equation, cf. eqns (5.3) and (5.4)?

(f) What can you say about the phase-space symmetry of a parametric oscillator with $\lambda > \lambda_{\rm th}$ initiated at $u = v = 0$ for short ($t_S \ll \tau_0$), intermediate ($\tau_0 \ll t_S \ll \tau_{\rm dwell}$), and long ($t_S \gg \tau_{\rm dwell}$) measurement times?

Tasks:

5.1 Open the code **Python Example 3** and study the noise-driven harmonic oscillator with a parametric drive $\lambda < \lambda_{\rm th}$ in a frame rotating at $\omega = \omega_0$. Use $\omega_0 = 1$, $Q = 100$, $m = 1$, $\beta = \eta = 0$, and $\varsigma_D = 1$ as default values. Calculate $\langle X^2 \rangle$ from the simulation result as a function of λ to estimate the fluctuation power, and compare to eqn (5.10). Does the result depend on ψ?

5.2 Add a resonant force F_0 with a phase $\theta = 0$ and vary the parametric pump with phase $\psi = -\pi/2$ between $\lambda = 0$ and $\lambda = \lambda_{\rm th}$. Compare the displacement standard deviations in the amplified and squeezed quadratures (with our choice of ψ, $q_+ = u$, and $q_- = v$). Calculate the ratio between the coherent response in the long-time limit and the noise standard deviation in the same direction. Does the ratio improve with parametric pumping? Do you note anything interesting regarding the timescales of the fluctuations in the two quadratures? Can you explain the observations?

5.3 Study the parametric oscillator above threshold, $\lambda > \lambda_{\rm th}$. Use $\omega_0 = 1$, $Q = 100$, $m = 1$, $\beta = 0.001$, $F_0 = 0$, $\varsigma_D = 0.01$, $\lambda = 1.5 \times \lambda_{\rm th}$, and an initial condition $(0, 0)$ as default values. You may need a shorter time step than with the harmonic oscillator to avoid numerical divergence. Run the simulation repeatedly and compare the results you obtain. Where do you see the effect of the force noise? What happens if you increase β?

5.4 With all parameters as above (and $\beta = 0.001$), initialize the system close to one of the phase states and observe the small fluctuations driven by ς_D. Find a way to determine the typical fluctuation frequency and chirality from the data. Can you use the method of characteristic exponent discussed in Sections 3.1.3 and 5.3.1 to verify your result?

5.5 Set $\beta = 0.01$ and test whether you can induce hopping between the phase states by increasing the force noise ς_D. Add a small external force to tilt the quasi-potential, such that the system spends, on average, more time in one phase state than in the other. Find a method to quantitatively evaluate the average dwell time of each phase state. Can you formulate a rule of thumb for the required sampling time t_{samp} and the total time t_S to obtain a reproducible result? What is the remaining statistical uncertainty of the average dwell time?

5.6 Change the detuning to obtain three attractors (the third at $u = v = 0$). Repeat the hopping experiment as a function of detuning and link the results to the Fokker–Planck solutions in Fig 5.11.

6

Coupled Harmonic Resonators

In previous chapters, we studied single resonators. We started in Chapter 1 with a harmonic oscillator whose Hamiltonian features only a quadratic potential term,

$$H = \frac{p^2}{2m} + \frac{1}{2}kx^2 \,. \tag{6.1}$$

Building up toward more complex systems, we then included higher-order potential terms, external driving and parametric pumping, and fluctuating forces in Chapters 2 to 4. In this chapter, we will study what happens when we couple several harmonic oscillators into a single system. We will focus on linear and energy-conserving coupling, moving from static coupling coefficients to time-dependent (parametric) coupling. Dissipative and nonlinear coupling will be discussed briefly toward the end of the chapter.

6.1 Static Coupling

A system of N linearly coupled harmonic oscillators with external forcing has the general Hamiltonian

$$H = \sum_i \left[\frac{p_i^2}{2m_i} + \frac{k_i}{2}x_i^2 + \sum_j J_{ij}x_i x_j - F_i(t)x_i \right] , \tag{6.2}$$

where each resonator i has position and momentum coordinates x_i and p_i, respectively, $k_i = m_i\omega_i^2$ are the spring constants, $J_{ij} = J_{ji}$ is the coupling coefficient between the resonators i and j,[1] and F_i are external forces. Including damping terms $\Gamma_i\dot{x}_i$ as in eqn (1.5), each of the resonators is subject to an EOM of the form

$$\ddot{x}_i + \Gamma_i\dot{x}_i + \frac{k_i}{m_i}x_i - \sum_j \frac{J_{ij}}{m_i}x_j = \frac{F_i}{m_i}\cos(\omega t + \theta_i) \,. \tag{6.3}$$

The system described by the coupled EOMs in eqn (6.3) can be cast into a matrix formulation similar to that introduced in Section 1.3 for a single resonator,

$$\dot{\mathbf{x}} = G\mathbf{x} + \mathbf{f}, \tag{6.4}$$

[1] An alternative way to write the coupling is with a quadratic force $\frac{1}{2}J_{ij}(x_i - x_j)^2$. This notation leads to the same form as eqn (6.2) after incorporating the terms $J_{ij}x_i^2$ and $J_{ij}x_j^2$ in the respective spring forces [31]. Such a procedure leads to a shift of the effective resonance frequencies relative to our notation but otherwise produces the same behavior, as we will observe in the following.

but we now identify $\mathbf{x}$ with a vector that contains all of the individual degrees of freedom, as well as their conjugate variables,

$$\mathbf{x} = \begin{bmatrix} x_1 \\ \dot{x}_1 \\ x_2 \\ \dot{x}_2 \\ \vdots \\ x_N \\ \dot{x}_N \end{bmatrix} . \tag{6.5}$$

The matrix G contains all of the terms governing the behavior of the individual resonators in its diagonal 2×2 blocks, while the off-diagonal blocks provide coupling between resonators,

$$G = \begin{bmatrix} 0 & 1 & 0 & 0 & \cdots & 0 & 0 \\ -\omega_1^2 & -\Gamma_1 & \frac{J_{12}}{m_1} & 0 & \cdots & \frac{J_{1N}}{m_1} & 0 \\ 0 & 0 & 0 & 1 & \cdots & 0 & 0 \\ \frac{J_{12}}{m_2} & 0 & -\omega_2^2 & -\Gamma_2 & \cdots & \frac{J_{2N}}{m_1} & 0 \\ \vdots & \vdots & \vdots & \vdots & \ddots & \vdots & \vdots \\ 0 & 0 & 0 & 0 & \cdots & 0 & 1 \\ \frac{J_{1N}}{m_N} & 0 & \frac{J_{2N}}{m_N} & 0 & \cdots & -\omega_N^2 & -\Gamma_N \end{bmatrix} . \tag{6.6}$$

The force vector is given as

$$\mathbf{f}(t) = \begin{bmatrix} 0 \\ \frac{F_1}{m_1} \cos(\omega t + \theta_1) \\ 0 \\ \frac{F_2}{m_2} \cos(\omega t + \theta_2) \\ \vdots \\ 0 \\ \frac{F_N}{m_N} \cos(\omega t + \theta_N) \end{bmatrix} . \tag{6.7}$$

We can split G into two parts as $G = G_0 + G_J$, of which the first contains only the diagonal 2×2 blocks that govern the dynamics of the uncoupled resonators, and the second one is responsible for the coupling. Applying the transfer matrix approach from eqn (1.28) with respect to the basis of G_0, we obtain

$$\mathbf{x}(t) = \Phi(t, 0)\mathbf{x}(0) + \int_0^t \Phi(t, t') \left[\mathbf{f}(t') + G_J \mathbf{x}(t')\right] dt' . \tag{6.8}$$

Since $\mathbf{x}$ also appears on the right-hand side of eqn (6.8), we cannot assume that each resonator is described in terms of its basis solutions $x_{a,b}$, cf. eqn (1.22). One way to deal with such a self-consistent equation is a perturbative treatment, as we have done during the averaging method in Chapter 2. In the present case, however, we note that

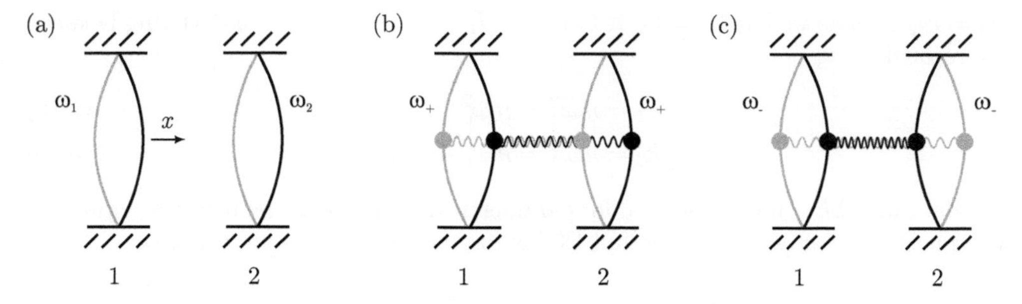

Fig. 6.1 (a) Two strings with similar natural frequencies that can be displaced along x. (b)–(c) In the presence of coupling between the strings, the stable solutions of the system consist of antisymmetrical and symmetrical normal modes with angular frequencies $\omega_\pm$.

the right-hand side of eqn (6.8) only couples to the left-hand side via linear off-diagonal terms. This allow us to find a transformation matrix U that produces a rotated matrix $\tilde{G} = U^{-1}GU$ whose off-diagonal 2×2 blocks contain only zeros.[2] Inserting $G = U\tilde{G}U^{-1}$ into eqn (6.4) and multiplying with U^{-1} from the left, we obtain

$$U^{-1}\dot{\mathbf{x}} = \tilde{G}U^{-1}\mathbf{x} + U^{-1}\mathbf{f}, \tag{6.9}$$

or, defining $\tilde{\mathbf{x}} \equiv U^{-1}\mathbf{x}$ and $\tilde{\mathbf{f}} \equiv U^{-1}\mathbf{f}$,

$$\dot{\tilde{\mathbf{x}}} = \tilde{G}\tilde{\mathbf{x}} + \tilde{\mathbf{f}}. \tag{6.10}$$

Note that $U^{-1}\dot{\mathbf{x}} = \dot{\tilde{\mathbf{x}}}$ because U is independent of time. From eqn (6.10), we see that $\tilde{G}$ acts on a rotated set of N degrees of freedom $\tilde{\mathbf{x}}$. We refer to these new degrees of freedom as **normal modes** of the coupled system. The normal modes are decoupled from each other, which means that we can modify the Wronskian matrix from eqn (1.22) to include the N normal mode variables,

$$\tilde{W}(t) = \begin{bmatrix} \tilde{x}_{1a}(t) & \tilde{x}_{1b}(t) & 0 & 0 & \cdots & 0 & 0 \\ \dot{\tilde{x}}_{1a}(t) & \dot{\tilde{x}}_{1b}(t) & 0 & 0 & \cdots & 0 & 0 \\ 0 & 0 & \tilde{x}_{2a}(t) & \tilde{x}_{2b}(t) & \cdots & 0 & 0 \\ 0 & 0 & \dot{\tilde{x}}_{2a}(t) & \dot{\tilde{x}}_{2b}(t) & \cdots & 0 & 0 \\ \vdots & \vdots & \vdots & \vdots & \ddots & \vdots & \vdots \\ 0 & 0 & 0 & 0 & \cdots & \tilde{x}_{Na}(t) & \tilde{x}_{Nb}(t) \\ 0 & 0 & 0 & 0 & \cdots & \dot{\tilde{x}}_{Na}(t) & \dot{\tilde{x}}_{Nb}(t) \end{bmatrix}, \tag{6.11}$$

and then apply eqns (1.28) and (1.29) for each normal mode separately to obtain a general solution for the coupled system.

6.1.1 The Normal Mode Picture for N = 2

In order to appreciate the processes underlying coupling phenomena, we consider the simple case of two nearly degenerate resonators 1 and 2, see Fig. 6.1(a). We assume

[2] It can be shown that this diagonalization is equivalent to a perturbative expansion to all orders.

$m_1 = m_2 \equiv m$ as well as $\Gamma_1 = \Gamma_2 \equiv \Gamma$, define $J_{12} = J$, and include a detuning between the resonators as

$$k_1 = m\omega_1^2 = m\omega_0^2 - \Delta k\,, \tag{6.12}$$
$$k_2 = m\omega_2^2 = m\omega_0^2 + \Delta k\,. \tag{6.13}$$

The detuning Δk will later be used for parametric driving and is defined in a symmetric way purely for convenience. The full EOMs for the coupled system are now

$$\ddot{x}_1 + \Gamma\dot{x}_1 + \frac{k_1}{m}x_1 - \frac{J}{m}x_2 = \frac{F_1}{m}\cos(\omega t + \theta_1)\,, \tag{6.14}$$
$$\ddot{x}_2 + \Gamma\dot{x}_2 + \frac{k_2}{m}x_2 - \frac{J}{m}x_1 = \frac{F_2}{m}\cos(\omega t + \theta_2)\,. \tag{6.15}$$

In our matrix notation, this simple system is written as

$$G = \begin{bmatrix} 0 & 1 & 0 & 0 \\ -\omega_0^2 + \frac{\Delta k}{m} & -\Gamma & \frac{J}{m} & 0 \\ 0 & 0 & 0 & 1 \\ \frac{J}{m} & 0 & -\omega_0^2 - \frac{\Delta k}{m} & -\Gamma \end{bmatrix}, \tag{6.16}$$

with the forcing term

$$\mathbf{f} = \begin{bmatrix} 0 \\ \frac{F_1}{m}\cos(\omega t + \theta_1) \\ 0 \\ \frac{F_2}{m}\cos(\omega t + \theta_2) \end{bmatrix}. \tag{6.17}$$

The transformation matrices that diagonalize this system are [31]

$$U^{-1} = \frac{1}{\sqrt{2}} \begin{bmatrix} 1 + \frac{\Delta k}{J}r_2 & 0 & r_2 & 0 \\ 0 & 1 + \frac{\Delta k}{J}r_2 & 0 & r_2 \\ 1 - \frac{\Delta k}{J}r_2 & 0 & -r_2 & 0 \\ 0 & 1 - \frac{\Delta k}{J}r_2 & 0 & -r_2 \end{bmatrix} \tag{6.18}$$

and

$$U = \frac{1}{\sqrt{2}} \begin{bmatrix} 1 & 0 & 1 & 0 \\ 0 & 1 & 0 & 1 \\ -\frac{\Delta k}{J} + r_1 & 0 & -\frac{\Delta k}{J} - r_1 & 0 \\ 0 & -\frac{\Delta k}{J} + r_1 & 0 & -\frac{\Delta k}{J} - r_1 \end{bmatrix}, \tag{6.19}$$

where we use the notation

$$r_1 \equiv \sqrt{1 + \frac{\Delta k^2}{J^2}}\,,$$
$$r_2 \equiv 1/r_1\,. \tag{6.20}$$

With this rotation, the rotated matrix $\tilde{G}$ takes on the desired form

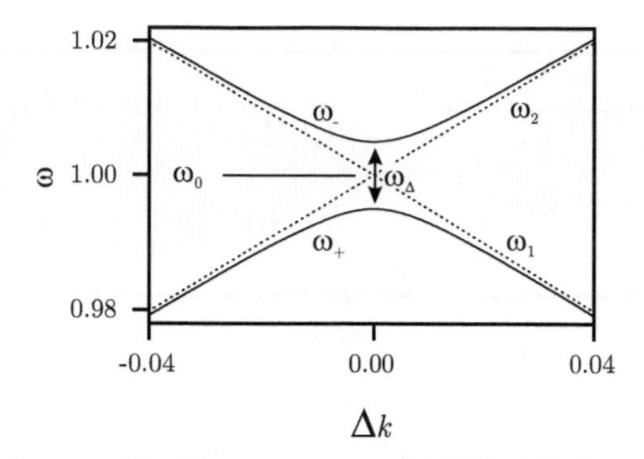

Fig. 6.2 Dependence of $\omega_\pm$ on finite detuning Δk in the spring constants for $\omega_{1,2} = 1$, $m = 1$, and $J = 0.01$. Dashed lines correspond to the solutions of $\omega_\pm$ for $J = 0$, which are identical to $\omega_{1,2}$. A horizontal line indicates ω_0.

$$\tilde{G} = \begin{bmatrix} 0 & 1 & 0 & 0 \\ -\omega_0^2 + r_1\frac{J}{m} & -\Gamma & 0 & 0 \\ 0 & 0 & 0 & 1 \\ 0 & 0 & -\omega_0^2 - r_1\frac{J}{m} & -\Gamma \end{bmatrix}, \tag{6.21}$$

and we obtain the rotated displacement variables

$$\tilde{\mathbf{x}} \equiv \begin{bmatrix} x_+ \\ \dot{x}_+ \\ x_- \\ \dot{x}_- \end{bmatrix} = \frac{1}{\sqrt{2}} \begin{bmatrix} (1 + \frac{\Delta k}{J} r_2)x_1 + r_2 x_2 \\ (1 + \frac{\Delta k}{J} r_2)\dot{x}_1 + r_2 \dot{x}_2 \\ (1 - \frac{\Delta k}{J} r_2)x_1 - r_2 x_2 \\ (1 - \frac{\Delta k}{J} r_2)\dot{x}_1 - r_2 \dot{x}_2 \end{bmatrix}, \tag{6.22}$$

We can see that with our choice of U, eqn (6.10) describes the evolution of two decoupled *hybrid* normal modes whose composition depends on the detuning Δk, see eqn (6.22). The corresponding frequencies $\omega_\pm$ are given by

$$\omega_\pm^2 = \omega_0^2 \mp \frac{1}{m}\sqrt{\Delta k^2 + J^2}, \tag{6.23}$$

and are plotted in Fig. 6.2 as a function of detuning. For large detuning, the normal modes approximate the uncoupled modes of the resonator (see dashed lines in Fig. 6.2). For small detuning, however, the normal modes are combinations of both resonators. For the usual case $J/m \ll \omega_0^2$, the minimum splitting between ω_- and ω_+ is equal to the angular exchange rate ω_Δ [31],

$$\omega_\Delta = \frac{2\pi}{t_\Delta} \approx \frac{J}{\omega_0 m}, \tag{6.24}$$

where t_Δ is the timescale for energy exchange between the resonators.

The meaning of t_Δ becomes clear from a simple thought experiment: let us assume the case of two resonators with $\Delta k = 0$, $F_{1,2} = \Gamma = 0$, $J \ll k_{1,2}$, and with the initial conditions $x_1 = 1$ and $x_2 = \dot{x}_1 = \dot{x}_2 = 0$, see Fig. 6.3. Releasing the system causes

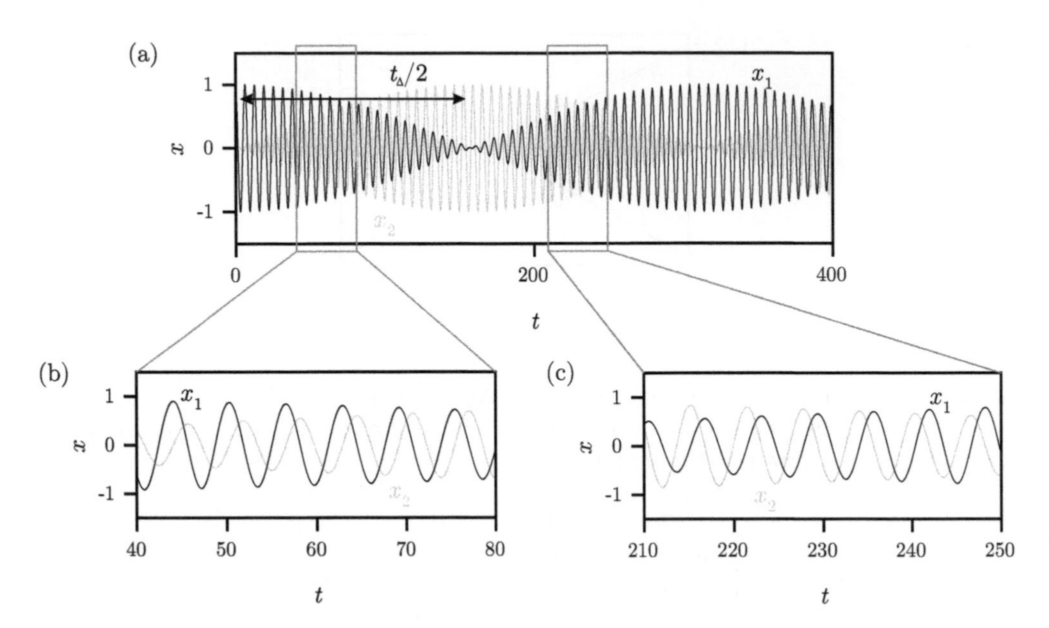

Fig. 6.3 (a) Numerical simulation of beating between two harmonic oscillators with initial conditions $x_1 = 1$ and $x_2 = \dot{x}_1 = \dot{x}_2 = 0$, and with the parameters $\omega_0 = 1$, $m = 1$, $\Gamma = 0$, $F_{1,2} = 0$, and $J = 0.02$. Black and gray lines correspond to x_1 and x_2, respectively. (b) Zoomed view showing the phase lag of $\pi/2$ between the driving resonator 1 and the driven resonator 2. (c) The roles of the resonators, and therefore the sign of the phase lag, is inverted at a later time during the beating cycle.

an initial oscillation $x_1 = e^{i\omega_0 t}$ that drives resonator 2 through the term $-\frac{J}{m}x_1$ in eqn (6.15). In response to this force, resonator 2 starts oscillating with a phase lag as $x_2 \propto e^{i(\omega_0 t - \pi/2)}$. The coupling term $-\frac{J}{m}x_2$ in eqn (6.14), which initially was zero, now grows in proportion to the amplitude of x_2 and drives resonator 1 in return. The response of resonator 1 has a phase lag relative to the driving force $-\frac{J}{m}x_2$, resulting in a motion $\propto e^{i(\omega_0 t - \pi)}$ that is opposed to the initial oscillation $x_1 = e^{i\omega_0 t}$. We therefore see that as resonator 1 drives resonator 2 to larger amplitudes, resonator 2 damps resonator 1 until $x_1 = 0$ and $x_2 = 1$ at a time $t_\Delta/2 = \frac{\pi m \omega_0}{J}$. At this point, the roles are exchanged and resonator 2 drives resonator 1, with the phase of resonator 1 being inverted. After a time span equal to t_Δ, all of the energy has returned to resonator 1, and the phase of resonator 2 is inverted as well. After $2t_\Delta$, the initial conditions (including phases) are restored. The energy exchange between the resonators leads to a beating phenomenon with amplitude modulations,

$$\cos(\omega_0 t)\cos\left(\frac{\omega_\Delta}{2}t\right) = \frac{1}{2}\cos\left(\left[\omega_0 + \frac{\omega_\Delta}{2}\right]t\right) + \frac{1}{2}\cos\left(\left[\omega_0 - \frac{\omega_\Delta}{2}\right]t\right). \tag{6.25}$$

These amplitude modulations produce the splitting of the original oscillation frequencies by ω_Δ.

A steady-state solution of the coupled system without beating can only be achieved for certain symmetric configurations, corresponding to the normal modes we found from the matrix formalism, cf. eqn (6.22). Since no coupling exists between these normal modes, no beating occurs when one of them is driven by an external force. In order to observe the beating, the energy decay via damping should be much slower than the energy exchange via the coupling, $\Gamma \ll \omega_\Delta$. This is the so-called **strong coupling limit**. The opposite case with $\Gamma > \omega_\Delta$ is referred to as the weak coupling regime.

6.1.2 Detuning and Normal Mode Coupling

From this point on, we choose as our basis the normal-mode solutions found for zero detuning, cf. Fig. 6.1(b) and (c). To distinguish this basis from the general notation $x_\pm(\Delta k)$, we introduce the new labels *symmetric* S and *antisymmetric* A for this basis,

$$\tilde{\mathbf{x}}(\Delta k = 0) = \begin{bmatrix} x_S \\ \dot{x}_S \\ x_A \\ \dot{x}_A \end{bmatrix} = \frac{1}{\sqrt{2}} \begin{bmatrix} x_1 + x_2 \\ \dot{x}_1 + \dot{x}_2 \\ x_1 - x_2 \\ \dot{x}_1 - \dot{x}_2 \end{bmatrix} . \tag{6.26}$$

The symmetric and antisymmetric variables $x_{S,A}$ are identical to $x_\pm$ for $\Delta k = 0$. The corresponding transformation matrix is

$$U(\Delta k = 0) = \frac{1}{\sqrt{2}} \begin{bmatrix} 1 & 0 & 1 & 0 \\ 0 & 1 & 0 & 1 \\ 1 & 0 & -1 & 0 \\ 0 & 1 & 0 & -1 \end{bmatrix} . \tag{6.27}$$

If we use $x_{S,A}$ as our new basis even in the presence of small detuning, we find that

$$\tilde{G} = \begin{bmatrix} 0 & 1 & 0 & 0 \\ -\omega_0^2 + \frac{J}{m} & -\Gamma & \frac{\Delta k}{m} & 0 \\ 0 & 0 & 0 & 1 \\ \frac{\Delta k}{m} & 0 & -\omega_0^2 - \frac{J}{m} & -\Gamma \end{bmatrix} . \tag{6.28}$$

As we can see from eqn (6.28), the normal modes defined for $\Delta k = 0$ become coupled for finite detuning $\Delta k \neq 0$. This tunable coupling between normal modes will provide us with a crucial tool for parametric coupling in Section 6.2. For now, rewriting eqns (6.14) and (6.15) in terms of $x_{S,A}$ with the help of eqn (6.28), we can define $\omega_{S,A}^2 = \omega_0^2 \pm \frac{J}{m}$ and obtain

$$\ddot{x}_S + \Gamma \dot{x}_S + \omega_S^2 x_S - \frac{\Delta k}{m} x_A = \frac{F_S}{m} \cos(\omega_S t + \theta_S) , \tag{6.29}$$

$$\ddot{x}_A + \Gamma \dot{x}_A + \omega_A^2 x_A - \frac{\Delta k}{m} x_S = \frac{F_A}{m} \cos(\omega_A t + \theta_A) , \tag{6.30}$$

where $F_{S,A}$ are components of $\tilde{f}$. Note that the detuning Δk acts on both normal modes equally, and that the roles of detuning and coupling have been exchanged with respect to eqns (6.14) and (6.15): here, it is the original coupling strength that determines the splitting between $\omega_{S,A}$, while a finite detuning Δk leads to coupling between the normal modes. In the following, we generate parametric coupling by modulating Δk as a function of time.

6.2 Nondegenerate Three-Wave Mixing

In the previous section, we saw how coupling between two resonance modes that are degenerate in frequency ($\omega_1 = \omega_2$) leads to nondegenerate normal modes ($\omega_S < \omega_A$). Through a shared detuning Δk, the normal modes can themselves be weakly coupled. In the following, we forget how the normal modes were formed and interpret eqns (6.29) and (6.30) as any two nondegenerate modes with amplitudes $X_{S,A}$. In contrast to the static coupling term J that we used for resonant coupling between degenerate modes, we allow for $\Delta k(t)$ to be time-dependent as $\Delta k(t) = 2\omega_0 g \cos(\omega_p t)$, where $2\omega_0 g$ is the modulation depth of the pump tone.[3]

For the nondegenerate modes that we consider here, energy exchange between the normal modes via a constant coupling term is off-resonant and thus inefficient. By contrast, for a time-dependent Δk, the coupling term for mode A or S appearing on the right-hand side of eqn (6.8) takes on the form

$$\begin{aligned}\frac{\Delta k}{m} x_{S,A} &= 2\omega_0 g \cos(\omega_p t + \phi_p) X_{S,A} \cos\left(\omega_{S,A} t + \phi_{S,A}\right) \\ &= \frac{2\omega_0 g X_{S,A}}{2} \left[\cos\left([\omega_{S,A} + \omega_p]\, t + \phi_{S,A}\right) + \cos\left([\omega_{S,A} - \omega_p]\, t + \phi_{S,A}\right)\right] , \end{aligned} \tag{6.31}$$

which can have components at the right frequencies for mutual driving if we choose the correct ω_p. In eqn (6.31) and in the following, we set the parametric phase offset to zero, $\phi_p = 0$, to allow for a succinct notation.

6.2.1 Driving at the Frequency Difference

From eqn (6.31), we see that a time-dependent modulation of the detuning creates sidebands in the coupling forces at the frequencies $\omega_{S,A} \pm \omega_p$. There are two important cases when these sidebands lead to significant coupling between nondegenerate modes. The first of these corresponds to driving at the **frequency difference**,

$$\omega_p = \omega_A - \omega_S , \tag{6.32}$$

as illustrated in Fig. 6.4(a). Here, the two modes drive each other via opposite sidebands, that is, the mode at ω_A drives the one at ω_S through the sideband at $\omega_A - \omega_p$, while in the opposite direction the upper sideband $\omega_S + \omega_p$ is relevant.

For an intuitive picture, imagine a case with starting conditions $x_S = 1$ and $x_A = 0$ and with $F_1 = \Gamma = 0$. In response to the coupling term

$$\frac{2\omega_0 g X_S}{2} \cos\left([\omega_S + \omega_p]\, t\right) , \tag{6.33}$$

[3] As we have seen in Section 6.1.2, for the special case when $x_{S,A}$ arise from two coupled degenerate modes $x_{1,2}$, Δk can be generated through a shared parametric pump, that is, an intentional detuning acting on $x_{1,2}$ simultaneously such that the normal base does not fully diagonalize G.

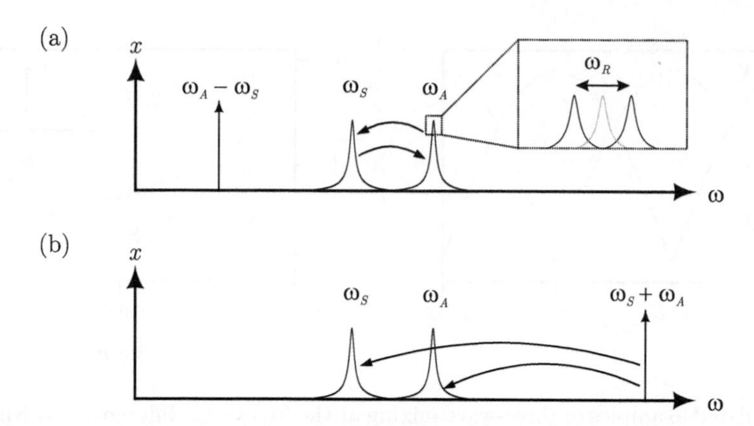

Fig. 6.4 Schematic representation of the frequencies involved in nondegenerate three-wave mixing. The insert shows how each normal mode splits into two peaks as a consequence of the periodic energy redistribution. (a) Driving at the frequency difference, $\omega_p = \omega_A - \omega_S$. (b) Driving at the frequency sum, $\omega_p = \omega_A + \omega_S$.

the upper mode at ω_A will begin to oscillate with a phase delay of $-\pi/2$, such that $x_A = X_A(t)\cos(\omega_A t - \pi/2)$. In turn, this will drive the lower mode through the coupling term

$$\frac{2\omega_0 g X_A}{2}\cos\left([\omega_A - \omega_p]\,t - \pi/2\right)\,, \tag{6.34}$$

which exerts a damping effect on x_S. In analogy to the case of two degenerate modes with a constant coupling coefficient in Section 6.1.1, x_S will drive x_A (and x_A will damp x_S) until all energy is stored on the upper mode, then x_S will increase again will opposite phase and so on. A numerical simulation of this dynamics is shown in Fig. 6.5(a).

In order to arrive at a more complete understanding of the coupled dynamics, we rewrite eqns (6.29) and (6.30) with $F_1 = 0$ in the form

$$\left[\frac{d^2}{dt^2} + \Gamma\frac{d}{dt}\right]\begin{bmatrix} x_S \\ x_A \end{bmatrix} + \begin{bmatrix} \omega_S^2 & -2\omega_0 g\cos(\omega_p t) \\ -2\omega_0 g\cos(\omega_p t) & \omega_A^2 \end{bmatrix}\begin{bmatrix} x_S \\ x_A \end{bmatrix} = 0\,. \tag{6.35}$$

We now apply the averaging method to first order, as described in Chapter 2. Here, however, we rotate the two equations of motion at two different frequencies, $\Omega_S = \omega_S + \delta_S$ and $\Omega_A = \omega_A + \delta_A$, where we introduce the small detuning parameters $\delta_{S,A}$. We also take into account the possibility that the parametric drive is detuned from the nominal frequency difference, $\omega_p = (\omega_A - \omega_S) + \delta_p$. In our averaging method, we regard Γ, g, $\delta_{S,A}$, and δ_p as parameters of order ϵ. This procedure yields, to first order in ϵ, the coupled slow-flow equations [31, 110]

$$\begin{bmatrix} \dot{u}_S \\ \dot{v}_S \\ \dot{u}_A \\ \dot{v}_A \end{bmatrix} = \frac{1}{2}\begin{bmatrix} -\Gamma & 2\delta_S & 0 & g \\ -2\delta_S & -\Gamma & -g & 0 \\ 0 & g & -\Gamma & 2\delta_A \\ -g & 0 & -2\delta_A & -\Gamma \end{bmatrix}\begin{bmatrix} u_S \\ v_S \\ u_A \\ v_A \end{bmatrix} \equiv G_{\text{diff}}\begin{bmatrix} u_S \\ v_S \\ u_A \\ v_A \end{bmatrix}\,, \tag{6.36}$$

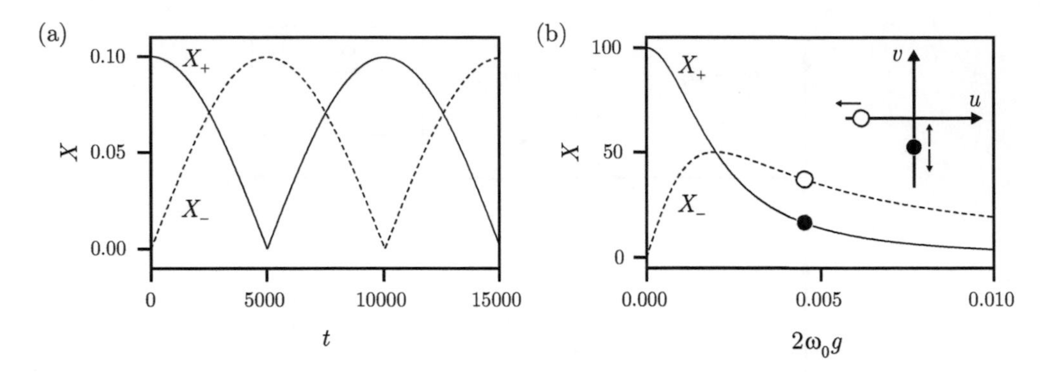

Fig. 6.5 Simulated examples of three-wave mixing at the frequency difference. (a) Numerical time evolution of eqn (6.38) with $\omega_S = 0.99$, $\omega_A = 1.01$, $m = 1$, $\Gamma = 10^{-6}$, $2\omega_0 g = 0.00125$, $\delta_i = 0$, and $F_{S,A} = 0$. We obtain $t_\Delta = 10053$. (b) Long-time response as a function of g obtained from eqn (6.42) with $\omega_S = 0.99$, $\omega_A = 1.01$, $m = 1$, $\Gamma = 10^{-3}$, $\delta_i = 0$, $F_S = 0.1$, and $F_A = 0$. Black and white spheres represent the amplitudes of the two modes for a particular value of $2\omega_0 g$ for comparison with the phase space graph in the inset. Small arrows show the different force contributions discussed in the text.

with $\omega_0 = \frac{\omega_S+\omega_A}{2}$ as before, and using the symbol G_{diff} for the propagator matrix. Writing the averaged degrees of freedom as

$$\mathbf{X} = \begin{bmatrix} u_S \\ v_S \\ u_A \\ v_A \end{bmatrix}, \tag{6.37}$$

we arrive at the form

$$\dot{\mathbf{X}} = G_{\text{diff}} \mathbf{X} \,. \tag{6.38}$$

An analysis of the entries of G_{diff} in eqn (6.36) is useful to understand the impact of the various detuning parameters. We can see that, to first order in ϵ, the detuning δ_p of the parametric drive does not appear at all. We only find that a detuning of the rotating frame from a mode, δ_i, leads to a weak coupling between the slow-flow quadratures u_i and v_i, as we should expect.

Equation (6.38) resembles the Bloch equations. Indeed, the energy exchange between x_S and x_A can be understood as a classical analogue to Rabi oscillations with bosonic modes [17, 29–31, 110, 111]. The corresponding quasi-Rabi frequency is the energy exchange rate between the modes, that is, the inverse of the time t_Δ that is required to transfer all energy from one mode to the other and back. For zero detuning, it is given by

$$\omega_R = \frac{2\pi}{t_\Delta} = g \,. \tag{6.39}$$

The frequency ω_R appears as an additional splitting of the normal modes into pairs of resonance peaks, as shown in the inset to Fig. 6.4(a). The frequency splitting that we see there corresponds to the amplitude spectrum of a single resonator. The energy cycling time t_Δ results in a gap of ω_R, in analogy to ω_Δ for the case of static coupling between degenerate resonators. However, in contrast to eqn (6.24), the quasi-Rabi frequency in eqn (6.39) is not equal to $J/m\omega_0$ but to g, which is only half of the coupling coefficient in eqn (6.35) divided by ω_0. This reduction of the energy exchange rate is due to the fact that only one of the two sidebands in eqn (6.31) is at the correct frequency to produce near-resonant driving, while the other half of the three-wave pump signal is off-resonant and can usually be neglected. This procedure, however, is only approximately correct for $\omega_R \ll |\omega_A - \omega_S|$. When $\omega_R \approx |\omega_A - \omega_S|$, the splitting seen in Fig. 6.4(a) is strong enough to bridge the gap between the modes [112]. In this **ultrastrong coupling regime**, additional effects appear. We will not further consider this special case in the main text, but will observe the corresponding effects in one of the exercises.

We can add a force term to eqn (6.38) as follows: we start with the vector **f** from eqn (6.17), transform it into its normal-mode form $\tilde{\mathbf{f}}$ as in eqn (6.10), and then apply the averaging method as in previous chapters, see for example eqns (5.1) and (5.2). For the typical case $\left|\frac{\omega_{S,A}}{\omega_0}\right| - 1 \ll 1$, we arrive at the simple notation

$$\mathbf{F} = \frac{1}{2m\omega_0}\begin{bmatrix} F_S\sin(\theta_S) \\ -F_S\cos(\theta_S) \\ F_A\sin(\theta_A) \\ -F_A\cos(\theta_A) \end{bmatrix} \tag{6.40}$$

and modify eqn (6.38) to become

$$\dot{\mathbf{X}} = G_{\text{diff}}\mathbf{X} + \mathbf{F}\,. \tag{6.41}$$

In this case, the system reaches a long-time limit response that depends on the coupling strength. To calculate these responses, we set $\dot{\mathbf{X}} = 0$ in eqn (6.41) and solve

$$\mathbf{X} = -G_{\text{diff}}^{-1}\mathbf{F}\,. \tag{6.42}$$

In Fig. 6.5(b), the long-time limit responses of X_S and X_A are displayed for the case $F_S \neq 0$. Here, all phases are fixed by the external force. The fact that both amplitudes are reduced with increasing g (beyond $g > \Gamma$) can alternatively be understood in the following simple way: starting from $g = 0$ and $\delta_i = 0$, a given external force F_S with $\theta_S = 0$ drives the symmetric mode to a long-time limit response

$$X_S = \frac{F_S}{m\omega_0\Gamma}e^{-\frac{\pi}{2}} = -i\frac{F_S}{m\omega_0\Gamma}\,, \tag{6.43}$$

which is simply the linear resonant response with a phase lag of $-\pi/2$. Regarding now the coupling term $\frac{\Delta k}{m}x_S$ with the amplitude $2\omega_0 g X_S$ as a driving force acting on x_A, we obtain via the same mechanism

$$X_A = \frac{gX_S}{\Gamma}e^{-\frac{\pi}{2}} = -i\frac{gX_S}{\Gamma}\,, \tag{6.44}$$

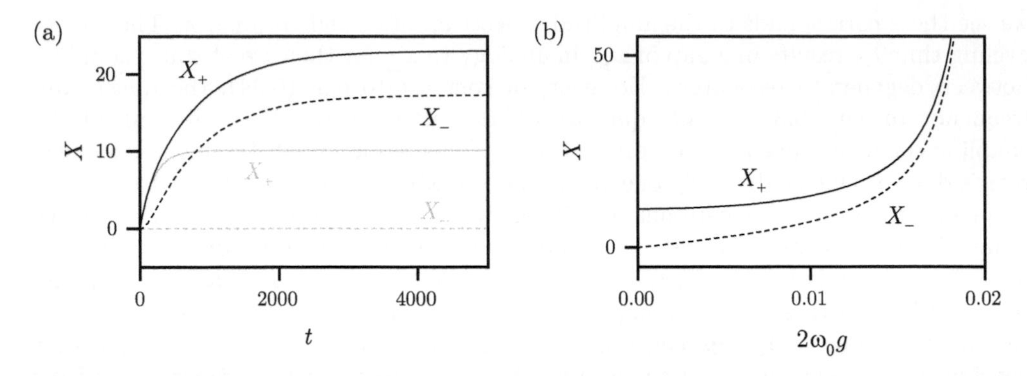

Fig. 6.6 Simulated examples of three-wave mixing at the frequency sum. (a) Numerical time evolution of eqn (6.41) with $\omega_S = 0.99$, $\omega_A = 1.01$, $m = 1$, $\Gamma = 10^{-2}$, $2\omega_0 g = 0$ (gray) and $g = 0.015$ (black), $\delta = 0$, $F_A = 0$, and $F_S = 0.1$. (b) Long-time response as a function of $2\omega_0 g$ obtained from eqn (6.42) with the same parameters as in (a).

which implies that the amplitude ratio $\frac{X_A}{X_S}$ increases linearly with the parametric driving strength $2\omega_0 g$. In turn, x_A exerts a coupling force $2\omega_0 g X_A/2$ back on x_S. The resulting steady-state equation for X_S is

$$X_S = -i\frac{F_S}{m\omega_0\Gamma} - i\frac{gX_A}{\Gamma} = -i\frac{F_S}{m\omega_0\Gamma} - \left(\frac{g}{\Gamma}\right)^2 X_S\,, \tag{6.45}$$

which we can solve for X_S to yield

$$X_S = -i\frac{F_S}{m\omega_0\Gamma}\left[1 + \left(\frac{g}{\Gamma}\right)^2\right]^{-1}. \tag{6.46}$$

As we see from eqn (6.46), X_S decreases approximately quadratically with g for $g \gg \Gamma$. Intuitively, strong coupling $2\omega_0 g$ loops the external force through X_A back to X_S with an inverted sign, generating a feedback damping force that approaches the strength of the original force F_S for $g \to \infty$. We emphasize that no energy conservation is expected in a driven and damped system.

6.2.2 Driving at the Frequency Sum

The second case where sidebands generate significant coupling arises for driving at the **frequency sum**, such that

$$\omega_p = \omega_A + \omega_S\,. \tag{6.47}$$

This case is illustrated in Fig. 6.4(b). As in the case of driving at the frequency difference, we can gain an understanding of the mechanics by studying the frequency components of the coupling forces, see eqn (6.31). When driving at the frequency sum, each mode couples to the other at the frequency difference $\omega_p - \omega_{S,A}$. In eqn (6.41), this symmetry leads to the same sign for all off-diagonal coupling elements, and we obtain with the averaging method [110]

$$\begin{bmatrix} \dot{u}_S \\ \dot{v}_S \\ \dot{u}_A \\ \dot{v}_A \end{bmatrix} = \frac{1}{2} \begin{bmatrix} -\Gamma & 2\delta_S & 0 & -g \\ -2\delta_S & -\Gamma & -g & 0 \\ 0 & -g & -\Gamma & 2\delta_A \\ -g & 0 & -2\delta_A & -\Gamma \end{bmatrix} \begin{bmatrix} u_S \\ v_S \\ u_A \\ v_A \end{bmatrix} + \mathbf{F} \tag{6.48}$$

which we write in a more compact way as

$$\dot{\mathbf{X}} = G_{\text{sum}}\mathbf{X} + \mathbf{F} \, . \tag{6.49}$$

In the numerical simulation in Fig. 6.6(a), we can identify two important features of this symmetrical driving. First, there is no beating of the amplitude as when driving at the frequency difference — the two modes drive each other simultaneously instead of alternately. Second, the time required to reach a saturation amplitude becomes longer for $g > 0$. This can be interpreted as an increased effective quality factor, similar to what we have studied for degenerate parametric driving in Section 3.1.1. In a similar fashion to the squeezing we noted in Section 5.2, we now observe a **two-mode squeezing** applied to both modes that corresponds to an effective gain [23, 113]

$$g_{\text{gain}} = \left[\left(1 + m\omega_0 g \sqrt{\frac{Q_S Q_A}{k_S k_A}} \right) \left(1 - m\omega_0 g \sqrt{\frac{Q_S Q_A}{k_S k_A}} \right) \right]^{-1} \tag{6.50}$$

with $Q_{S,A}$ and $k_{S,A}$ the quality factors and spring constants of the two modes, respectively. Since we assume $Q_S = Q_A$ and $k_{S,A} \approx m\omega_0^2$, we find that g_{gain} diverges for

$$\frac{g}{\omega_0^2} \geq \frac{2}{Q}, \tag{6.51}$$

which is analogous to the threshold for parametric instability in the case of degenerate three-wave mixing (replacing $\omega_0\lambda \longrightarrow 2g$). The main advantage of the nondegenerate case is that the two resonator phases are not fixed, causing the amplification to be phase insensitive. Instead, the sum of the two phases is locked when the modulation exceeds the parametric threshold [114, 115].

We can calculate the long-time limit response of the two mode amplitudes in the regime below threshold from

$$\mathbf{X} = -G_{\text{sum}}^{-1}\mathbf{F} \, . \tag{6.52}$$

An example calculated with eqn (6.52) is shown in Fig. 6.6(b). The amplitudes diverge towards $2\omega_0 g = 0.02$, in agreement with the threshold obtained from eqn (6.51).

6.3 Alternative Types of Coupling

So far, we have limited the discussion to linear and energy-conserving coupling mechanisms that can be expressed with Hamiltonian terms proportional to $(x_1 - x_2)^2$ or

x_1x_2. Here, we will briefly discuss alternative forms of coupling, with the aim to allow for a direct comparison of various effects found in literature. We will keep these discussions simple and brief, since we will not need the corresponding effects in the following chapters.

6.3.1 Dissipative Coupling

A first case that we investigate is **dissipative coupling**, corresponding to two equations of motion with coupling terms proportional to $\dot{x}_i$ [116],

$$\ddot{x}_1 + \Gamma\dot{x}_1 + \omega_1^2 x_1 - \frac{J_1}{m}\dot{x}_2 = 0\,, \tag{6.53}$$

$$\ddot{x}_2 + \Gamma\dot{x}_2 + \omega_2^2 x_2 - \frac{J_2}{m}\dot{x}_1 = 0\,. \tag{6.54}$$

In order to understand what dynamics arises from such coupling, let us examine the same example as in Section 6.1.1. Namely, we set $\omega_1 = \omega_2$ and $J_1 = J_2 > 0$, and start from $x_1 = 1$ and $x_2 = \dot{x}_1 = \dot{x}_2 = 0$. As before, we have an initial oscillation $x_1 = X_1 e^{i\omega_0 t}$ with X_1 designating the resonator's real amplitude. The effective coupling force $\frac{J_2}{m}\dot{x}_1$ acting on resonator 2 has a phase delay of $-\pi/2$ relative to x_1, leading to an oscillation $x_2 = X_2 e^{i\omega_0 t}$, that is, in phase with x_1. As x_2 grows, it leads to a force acting on x_1 through the term $\frac{J_1}{m}\dot{x}_2$. Again, the driven motion of x_1 is in phase with x_2. In this setup, therefore, the two resonators keep driving each other in phase. If the mutual driving is weaker than the damping forces of each resonator, $\Gamma > J_{1,2}$, the amplitudes $X_{1,2}$ tend to become equal and decrease together, as shown in gray in Fig. 6.7(a). In the opposite case, $\Gamma < J_{1,2}$, the amplitudes diverge as a consequence of the mutual driving, as shown in black for comparison.

Changing the signs of $J_{1,2}$ inverts the phase relationship between the resonators. For $J_1 = J_2 < 0$, the two resonators drive each other with a fixed phase difference of π. However, the mutual driving still results in unlimited growth if $J_{1,2} > \Gamma$. We conclude that symmetric dissipative coupling is not an energy-conserving process and requires active addition of energy, for instance via a signal amplifier [58].

Dissipative coupling becomes an energy-conserving process when the coupling coefficients have opposite signs, $J_1 = -J_2$. In this case, the slowly growing oscillation of x_2 drives x_1 with a phase that opposes that of the initial condition, leading to a conservative, periodic energy exchange between the two resonators. During the exchange cycle, the relative phase between the resonators switches back and forth between 0 and π. A numerical simulation of this case (with negligible damping Γ) is shown in Fig. 6.7(b). Note that this is not identical to the dynamics treated in Section 6.1.1 and displayed in Fig. 6.3, where the two resonators had phase differences of $\pm\pi/2$. Nevertheless, the same spectral splitting is observed due to the beating. From a comparison of the coupling terms in eqns (6.53) and (6.54) with those in eqns (6.14) and (6.15), we realize that the coupling coefficients in the former have units of 1/s, not 1/s^2 as in Section 6.1.1, and that the angular splitting must now be described by

$$\omega_\Delta \approx \frac{J_{1,2}}{\omega_{1,2}^2 m}\,. \tag{6.55}$$

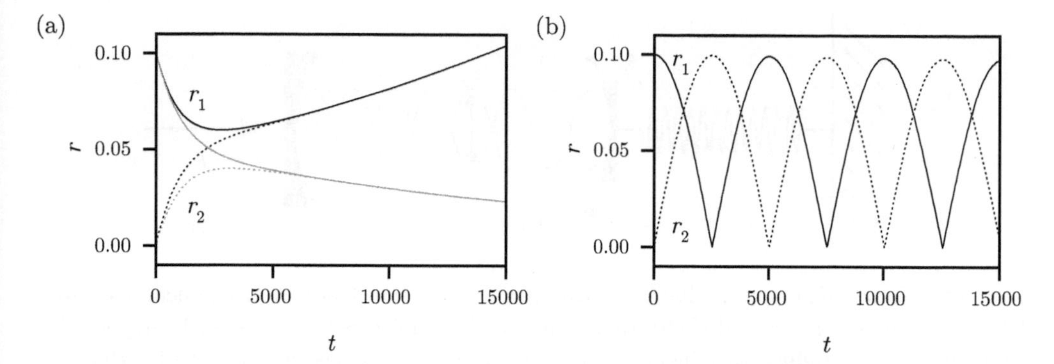

Fig. 6.7 Simulated examples of dissipative coupling with a pair of degenerate resonators without drive, with initial conditions $x_1 = 1$ and $x_2 = \dot{x}_1 = \dot{x}_2 = 0$. (a) $\omega_1 = \omega_2 = 1$, $m = 1$, $\Gamma = 10^{-3}$, $J_1 = J_2 = 0.9 \times 10^{-3}$ (gray), and $J_1 = J_2 = 1.1 \times 10^{-3}$ (black). Solid and dashed lines indicate the amplitudes of resonator 1 and 2, respectively. (b) The same simulation for $\Gamma = 10^{-6}$ and $J_1 = -J_2 = 1.25 \times 10^{-3}$. Note that the energy exchange is two times faster than in Fig. 6.5(a).

For the simulation in Fig. 6.7(b), this amounts to a time $t_\Delta = 2\pi/\omega_\Delta \approx 5027$ for a full energy exchange cycle, in agreement with the numerical results.

6.3.2 Nonlinear Coupling

To conclude our survey of mode coupling, we take a glance at energy-conserving, nonlinear coupling, as described by the Hamiltonian

$$H = \frac{p_1^2}{2m_1} + \frac{p_2^2}{2m_2} + \frac{1}{2}m_1\omega_1^2 x_1^2 + \frac{1}{2}m_2\omega_2^2 x_2^2 + J_{nl}x_1x_2^2 - F_d\cos(\omega_d t)\,x_2. \tag{6.56}$$

This Hamiltonian is important for the field of cavity optomechanics, where a mechanical mode at ω_1 and an optical cavity mode at $\omega_2 \gg \omega_1$ are nonlinearly coupled through the coefficient J_{nl}. A comprehensive derivation and analysis of such systems is presented in Ref. [33]. In this section, we provide an overview of their basic features with a simple treatment.

We transform the Hamitonian in eqn (6.56) into two coupled differential equations,

$$\ddot{x}_1 + \Gamma_1 x_1 + \omega_1^2 x_1 + J_{nl}x_2^2 = 0\,, \tag{6.57}$$

$$\ddot{x}_2 + \Gamma_2 x_2 + \omega_2^2 x_2 + J_{nl}x_1x_2 = F_d\cos(\omega_d t)/m_2\,. \tag{6.58}$$

We see that x_1 provides a parametric modulation for resonator 2, while resonator 2 acts as a nonlinear force onto x_1. A canonical experimental setup for this situation is shown in Fig. 6.8.

First, we assume that the "optical" resonator 2 is driven at $\omega_d = \omega_2$ by an external force with amplitude F_d, while the "mechanical" resonator 1 with $\omega_1 \ll \omega_2$ is oscillating as $X_1\cos(\omega_1 t + \phi_1)$, for instance under the influence of thermal force noise.[4]

[4] We adopt the terms used for mechanical systems for both modes, being aware that, for example, the term *mass* is not directly applicable to an optical mode.

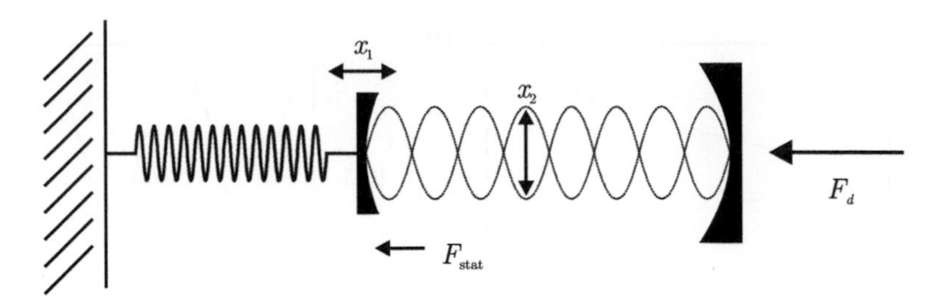

Fig. 6.8 Sketch of an optomechanical setup that realizes eqn (6.56). A mechanical resonator 1 forms one end of an optical cavity that consists of two mirrors (concave black shapes). An electromagnetic standing wave inside the cavity is indicated by dashed lines and is driven by a force field F_d. A displacement x_1 changes the length of the cavity, leading to a modulation of ω_2. The power circulating in the cavity (proportional to x_2^2) exerts an optical pressure onto the movable mirror and thus provides a force for x_1.

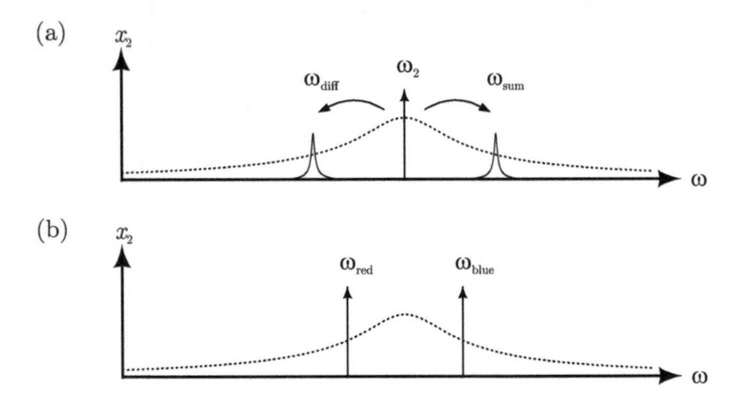

Fig. 6.9 (a) Schematic illustration of the "optical" power spectrum of resonator 2 (dashed line) and the sidebands generated at $\omega_{\text{diff}} = \omega_2 - \omega_1$ and $\omega_{\text{sum}} = \omega_2 + \omega_1$ by oscillations of x_1. (b) Negative (*red*) and positive (*blue*) detuning of ω_d from ω_2 lead to positive and negative feedback damping Γ_{FB} of resonator 1, respectively.

In the limit $J_{nl}x_1 \ll \omega_2^2$, a displacement x_1 instantaneously shifts ω_2 by $\frac{J_{nl}x_1}{2\omega_2} \ll \omega_2$. We obtain for resonator 2 the typical time trace of a frequency-modulated harmonic oscillator [117]

$$x_2 = X_2 \cos\left(\omega_2 t + \frac{\Delta_{nl}}{\omega_1}\cos\left(\omega_1 t + \phi_1\right)\right) . \tag{6.59}$$

For $\frac{\Delta_{nl}}{\omega_1} \ll 1$ and in the long-time limit, the frequency modulation leads to sidebands in the spectrum of resonator 2 at the frequencies $\omega_{\text{sum}} = \omega_2 + \omega_1$ and $\omega_{\text{diff}} = \omega_2 - \omega_1$, with amplitudes $\frac{\Delta_{nl}}{2\omega_1}$ relative to that of the peak at ω_2, see Fig. 6.9 [117]. These side peaks can be used to read out the oscillations of x_1 with high precision [33].

While the oscillation of x_1 generates frequency sidebands at $\omega_2 \pm \omega_1$, the oscillation of x_2 also exerts an effect on resonator 1 through the coupling term $J_{nl}x_2^2$ [118]. This force comprises a static component $F_{\text{stat}}/m_1 = J_{nl}X_2^2/2$ that corresponds to the photon pressure in an optomechanical setup as shown in Fig. 6.8, that is, the photon recoil at the left mirror provides an approximately static force. When resonator 1 is at rest with a deflection x_1 and resonator 2 is driven on resonance at $\omega_d = \omega_2$, the amplitude X_2 depends on x_1 via the detuning $\Delta_{nl} = \frac{J_{nl}x_1}{2\omega_2}$ as

$$X_2(\Delta_{nl}) = \frac{F_d/m_2}{\sqrt{\left(\omega_2^2 - \Delta_{nl}^2\right)^2 + \Delta_{nl}^2\Gamma_2^2}} . \tag{6.60}$$

For a given amplitude X_2, the differentiation of the static force with respect to x_1 leads in turn to an optical spring and a frequency shift of resonator 1,

$$\omega_{1,\text{opt}} = \left(\omega_1^2 - \frac{d}{dx_1}\frac{F_{\text{stat}}}{m_1}\right) . \tag{6.61}$$

Importantly, X_2 in eqn (6.60) responds to changes in Δ_{nl} (and therefore x_1) with a time delay. This delay is nothing other than the decay time $\tau_2 = 2Q_2/\omega_2$ that governs the energy transfer between resonator 2 and its environment, as we have seen, for example, in Fig. 1.4. Resonator 2 requires a finite time to adapt its amplitude to changes in the detuning. As a consequence, resonator 1 experiences a retarded feedback force $J_{nl}X_2^2(t)/2 = F_{\text{FB}}(t)/m_1$ in response to its own displacement changes. We write $F_{\text{FB}}(t)/m_1 = \omega_{\text{FB}}^2 x(t) + \Gamma_{\text{FB}}\dot{x}(t)$, by which we mean that the force can be split into an *in-phase component* that follows x instantaneously, and an *out-of-phase component* that follows with a phase delay of $\pi/2$ and is therefore proportional to $\dot{x}$. Inserting these force terms into eqn (6.57), we understand that $\omega_{\text{FB}}^2 x$ leads to an additional frequency shift, while $\Gamma_{\text{FB}}\dot{x}$ provides a damping effect.

The sign and strength of ω_{FB}^2 and Γ_{FB} depend on whether an increase of x_1 leads to an increase or decrease of X_1, see eqn (6.60). What makes this coupled system so useful is that the detuning of ω_2 from ω_d can be controlled by tuning the driving frequency (i.e. the laser wavelength in typical experiments). Driving resonator 2 below (above) its resonance frequency gives rise to a negative (positive) spring force ω_{FB}^2 and a positive (negative) damping coefficient Γ_{FB}. This **sideband control** is a powerful tool to decrease the effective temperature, and thereby the thermal fluctuations, of mechanical resonators through interaction with an optical or microwave cavity [33].

Chapter summary

- In Chapter 6, we move to systems of coupled harmonic oscillators.
- We start with the case of N degenerate damped, harmonic oscillators with linear coupling, cf. eqn (6.2). The coupling leads to energy exchange between the resonators and to **beating** of the individual amplitudes, cf. Fig. 6.3.
- For particular combinations of the two resonator oscillators, stable amplitudes arise in the coupled system. These combinations are referred to as **normal modes**, cf. Figs. 6.1 and 6.2. Normal modes form a new basis of uncoupled, nondegenerate oscillation modes. For two identical resonators, we find symmetric and antisymmetric oscillations with displacement coordinates x_S and x_A, respectively, cf. eqns (6.29) and (6.30).
- The nondegenerate normal modes can be **coupled through parametric modulation** (three-wave mixing) of the original resonators, which creates a time-dependent coupling term Δk, cf. Fig. 6.4 and eqn (6.31). If the parametric pumping frequency is equal to the frequency difference between the normal modes, $\omega_p = \omega_A - \omega_S$, the dynamic coupling leads to peak splitting and energy exchange between the normal modes, cf. Fig. 6.6 and eqn (6.36). For $\omega_p = \omega_A - \omega_S$, the parametric coupling gives rise to mutual driving of the modes, cf. Fig. 6.6 and eqn (6.48).
- Instead of linear coupling between two harmonic oscillators, one can study dissipative coupling, cf. Fig. 6.7 and eqns (6.53) and (6.54). Here, symmetric coupling cannot conserve energy and leads to mutual driving. If the coupling terms are defined in an antisymmetric way, the coupling generates beating between the resonator amplitudes in a way that is similar (but not equivalent) to linear coupling.
- We also briefly mention nonlinear coupling, which has received an enormous amount of attention in the cavity optomechanics community, cf. Fig. 6.8 and eqns (6.57) and (6.58). In particular, driving the system with an external force at the sum of difference of the resonance frequencies can be used to engineer energy exchange or mutual driving between the oscillators, cf. Fig. 6.9. These phenomena are similar to those arising from the three-wave mixing discussed previously.

Exercises

Check questions:

(a) In a system of coupled harmonic oscillators, what is the fundamental difference between the original basis (resonator basis) and the normal-mode basis? Is there a fundamental difference between a normal mode and a single, uncoupled oscillator?

(b) When beating occurs, what is the reason why energy is transmitted first in one direction, then in the other? What determines the direction of the energy flow at any given moment?

(c) What are the spectral signatures (peaks) observed in a ringdown experiment and a frequency sweep driving with an external force?

(d) Why do physicists often exclaim that “two is much larger than one”?

Tasks:

6.1 Open the code **Python Example 6** and study two degenerate, coupled harmonic oscillators A and B. First, use $\omega_{A,B} = 1$, $Q_{A,B} = 10^6$, $m_{A,B} = 1$, $J = 0.5$, and $F_{A,B} = 0$ as default values. Let the system ring down from the initial condition $x_A(0) = 1$ and $x_B(0) = 0$ with a step size $dt = 0.1$ and a final time $t_S = 1000$. Compare the resulting beating frequency with the analytical result in eqn (6.24).

6.2 Change the initial conditions of the ringdown. Can you achieve that only the symmetric or antisymmetric modes are activated?

6.3 Change the properties of the resonators. What changes in the beating when the resonator frequencies or masses are not identical?

6.4 Set the initial conditions to $x_A(0) = 0$ and $x_B(0) = 0$. Using eqn (6.23), find out how to apply an external force that drives only the symmetric or antisymmetric mode.

6.5 Set the resonance frequencies to $\omega_A = 1.01$ and $\omega_B = 0.99$ and the coupling to $J = 0.001$ with $\omega_{\text{Diff}} = \omega_B - \omega_A$. Run a simulation over $t_S = 20000$ from initial conditions $x_A(0) = 1$ and $x_B(0) = 0$ without external forces. In addition to the slow energy exchange between the resonators, you should observe small wiggles on a timescale $\pi/|\omega_B - \omega_A|$. This is the effect of the second (off-resonant) sideband, which we usually neglect and which does not appear in the slow-flow solution in eqn (6.36). Increase J or decrease $|\omega_B - \omega_A|$ until $\omega_R \approx |\omega_A - \omega_S|$ and observe what happens here. Can you explain what you see?

6.6 Run a ringdown for a coupled system with the parameters $\omega_A = 1.01$, $\omega_B = 0.99$, $Q_{A,B} = 10^3$, $m_{A,B} = 1$, $J = 0.001$, and a pump at the frequency sum, $\omega_{\text{Diff}} = \omega_B + \omega_A$. Interpret the outcome after a sufficiently long time t_S with the help of eqn (10.7). Then run the simulation with an external force F_A applied at one of the resonance frequencies and try to observe a similar result.

7
Coupled Parametric Oscillators

In Chapter 6, we studied coupled harmonic oscillators. The most important lesson was that for linear, energy-conserving coupling, there usually exists a basis of uncoupled normal modes which allows us to apply our standard tools from Chapter 1 to the system. In this chapter, we want to extend the discussion to coupled nonlinear parametric oscillators. The nonlinearity does not allow us to find a generally valid normal-mode basis, and the interplay of nonlinearity and coupling can lead to situations with high complexity. Nevertheless, the intuition we draw from a normal-mode picture is still helpful to understand coupled parametric oscillators [52, 119, 120]. This is particularly true for the onset of linear instability, which we described as Arnold tongues for individual resonators. The Arnold tongues are defined by the effective damping of the oscillators at zero displacement, where nonlinearities can be neglected, cf. eqn (3.11). As we will see, coupling leads to the emergence of several Arnold tongues whose positions follow the underlying normal-mode structure of the network. As the exploration of parametric oscillator networks is a very active research field, we limit our discussion here to a few exemplary cases with static coupling.

7.1 Equations for N Coupled Parametric Oscillators

In contrast to our prior treatment of coupled harmonic resonators, we now include a parametric pump at a frequency $\omega_p = 2\omega$, nonlinear terms, and force noise as in Chapter 4. These additions yield N coupled differential equations of the form

$$\ddot{x}_i + \omega_i^2 \left[1 - \lambda \cos\left(2\omega t\right)\right] x_i + \beta_i x_i^3 + \Gamma_i \dot{x}_i + \eta_i x_i^2 \dot{x}_i - \sum_{j \neq i} J_{ij} x_j = \frac{\xi_i}{m}, \tag{7.1}$$

where ω_i is the eigenfrequency, Γ_i the dissipation rate, β_i the effective Duffing nonlinearity, η_i the nonlinear damping coefficient, and ξ_i the fluctuating force of the ith oscillator, and $J_{ij} = J_{ji}$ is the coupling term between resonators i and j with $i, j = 1, 2, \ldots, N$ for all $i \neq j$. We fixed all parametric pumping phases to $\psi_i = 0$ for simplicity, cf. eqn (3.1).

An illustration of a network of coupled parametric oscillators is shown in Fig. 7.1. We will in the following distinguish four main cases: the system can feature weak or strong coupling J on one hand, and weak or strong parametric pumping λ on the other. We will attempt a rough classification of the phenomena arising in parametric oscillator networks based on these two parameters in Section 7.2.

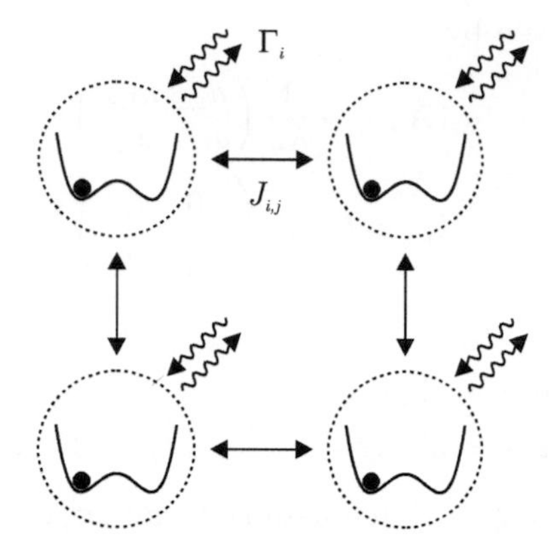

Fig. 7.1 Schematic representation of a network of four parametric oscillators with coupling coefficients J_{ij} and coupling to an environment characterized by the dissipation rates Γ_i.

7.1.1 Slow-Flow Formulation

For weakly nonlinear systems, a slow-flow formulation of N coupled resonators with frequencies ω_i $(i = 1, 2, ..., N)$ can be written in a frame rotating at $\omega_p/2 = \omega$ as:

$$\dot{\mathbf{X}} = G(\mathbf{X})\mathbf{X} + \mathbf{F}_{\text{noise}} \,. \tag{7.2}$$

Here,

$$\mathbf{X} = \begin{bmatrix} u_1 \\ v_1 \\ \vdots \\ u_N \\ v_N \end{bmatrix} \tag{7.3}$$

contains all of the slowly varying phase-space degrees of freedom of the resonators, while the term

$$\mathbf{F}_{\text{noise}} = \frac{1}{m} \begin{bmatrix} \Xi_{u,1} \\ \Xi_{v,1} \\ \vdots \\ \Xi_{u,N} \\ \Xi_{v,N} \end{bmatrix} \tag{7.4}$$

comprises the thermal force noise as in eqns (5.1) and (5.2). The rotating force noise terms $\Xi_{u,i}$ and Ξ_{v_i} are calculated as in eqn (4.26), m is the mass (assumed to be identical for all resonators), and the matrix G is [76]

$$G(\mathbf{X}) = \begin{bmatrix} a_1(\mathbf{X}) & b_{1,2} & \cdots & b_{1,N} \\ b_{2,1} & a_2(\mathbf{X}) & \ddots & \vdots \\ \vdots & \ddots & \ddots & b_{(N-1),N} \\ b_{N,1} & \cdots & b_{N,(N-1)} & a_N(\mathbf{X}) \end{bmatrix} . \tag{7.5}$$

The matrix entries are given by

$$a_i(\boldsymbol{X}) = -\frac{1}{8\omega}\begin{pmatrix} a_{i,1} & a_{i,+} \\ a_{i,-} & a_{i,1} \end{pmatrix} \tag{7.6}$$

$$b_{i,j} = b_{j,i} = \begin{pmatrix} 0 & \frac{J_{ij}}{2\omega} \\ -\frac{J_{ij}}{2\omega} & 0 \end{pmatrix} \tag{7.7}$$

with the terms

$$\begin{aligned} a_{i,1} &= 4\Gamma_i\omega + \eta_i\omega X_i^2 \\ a_{i,\pm} &= 2(\lambda\omega_i^2 + 2(\omega_i^2 - \omega^2)) \pm 3\beta_i X_i^2 \end{aligned} \tag{7.8}$$

where we have used $X_i^2 = u_i^2 + v_i^2$ and assumed that $\omega_i^2/\omega^2 - 1 \ll 1$ for all i.

7.2 Examples for $N = 2$

In the following, we study the interplay between parametric pumping and linear coupling in a system of two identical resonators with $\omega_{1,2} \equiv \omega_0$, $\Gamma_{1,2} \equiv \Gamma$, $\beta_{1,2} \equiv \beta$, $\eta_{1,2} \equiv 0$, and with coupling $J_{12} \equiv J$. The slow-flow equation from eqn (7.2) to (7.8) can then be simplified to [86]

$$\begin{aligned} \dot{u}_{1,2} &= -\frac{\Gamma u_{1,2}}{2} - \left(\frac{3\beta}{8\omega}X_{1,2}^2 + \frac{\omega_0^2 - \omega^2}{2\omega} + \frac{\lambda\omega_0^2}{4\omega}\right) v_{1,2} + \frac{J v_{2,1}}{2\omega} + \frac{\Xi_{u,1,2}}{m}, \\ \dot{v}_{1,2} &= -\frac{\Gamma v_{1,2}}{2} + \left(\frac{3\beta}{8\omega}X_{1,2}^2 + \frac{\omega_0^2 - \omega^2}{2\omega} - \frac{\lambda\omega_0^2}{4\omega}\right) u_{1,2} - \frac{J u_{2,1}}{2\omega} + \frac{\Xi_{v,1,2}}{m}. \end{aligned} \tag{7.9}$$

We divide the results into four cases that are characterized by parametric pumping below ($\lambda < \lambda_{\text{th}}$) and above ($\lambda > \lambda_{\text{th}}$) threshold, as well as strong versus weak coupling. The definition we use for *weak coupling* in this case relies on a comparison of the normal mode splitting ω_Δ (cf. Section 6.1.1) to the width of the Arnold tongue ω_{AT} in the absence of damping ($Q \to \infty$). From eqn (3.11), we know this width to be

$$\omega_{\text{AT}} = \omega_0 \left[\left(1 + \frac{\lambda}{2}\right)^{1/2} - \left(1 - \frac{\lambda}{2}\right)^{1/2}\right]. \tag{7.10}$$

To make our definition of weak coupling independent of λ, we choose to evaluate ω_{AT} at $\lambda = \lambda_{\text{th}}$ of the damped system. Using the approximation

$$\left(1 \pm \frac{1}{Q}\right)^{1/2} \approx 1 \pm \frac{1}{2Q}, \tag{7.11}$$

we find

$$\omega_{\text{AT}}(\lambda_{\text{th}}) = \Gamma, \tag{7.12}$$

recovering the definition of weak coupling $\omega_\Delta < \Gamma$ used commonly for harmonic oscillators. An overview of the case of two weakly coupled parametric oscillators is shown in Fig. 7.2, and the corresponding graph for the case of strong coupling is provided in Fig. 7.7.

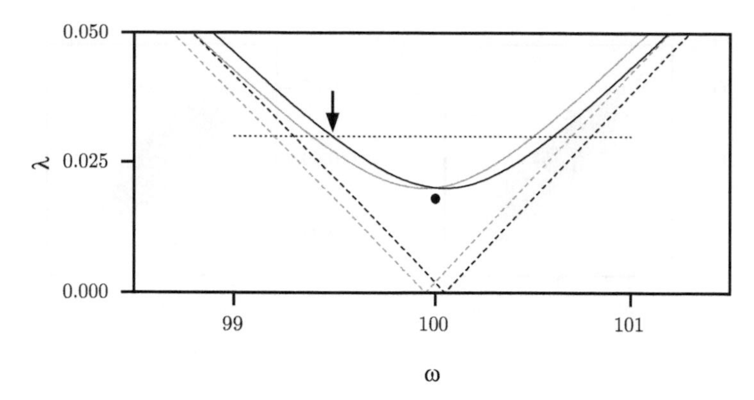

Fig. 7.2 Arnold tongues of a system of two weakly coupled degenerate parametric oscillators. Each resonator has $\omega_0 = 100$, $\Gamma = 1$, and $m = 1$. The static coupling between the resonators is $J = 10$, giving rise to a normal-mode splitting of $\omega_\Delta = \frac{J}{\omega_0 m} = 0.1$. Gray (black) lines correspond to the symmetric (antisymmetric) normal mode, while dashed (solid) lines indicate the boundaries of the Arnold tongue without (with) damping, cf. the discussion in Section 3.1.2 and Fig. 3.5. A black dot and a dotted horizontal line indicate the positions of the simulations shown in Figs. 7.3 and 7.4, respectively. The arrow indicates the position of the switch in Fig. 7.4.

7.2.1 Weak Coupling and Weak Pumping

In the limit of vanishing coupling and parametric pumping, each resonator is driven by thermal noise. We assume this force noise to be small, such that nonlinear terms in the equations of motion are negligible and we arrive at the same behavior as in Fig. 5.2(a). Increasing λ (but remaining below λ_{th}), both resonators are subject to classical squeezing as in Fig. 5.2(b). Adding weak coupling J leads to a slow beating between the resonators, or equivalently to normal modes with a small splitting ω_Δ. As the timescale of the noise-driven fluctuations is much faster than the beating dynamics, the latter is not prominent and the system is dominated by the effect of parametric squeezing close to ω_0. A rotation into a normal-mode picture does not reveal any new insight in this case, see Fig. 7.3. Close to the edge of the undamped Arnold tongue, the parametric amplification becomes gradually weaker and the system smoothly reverts to the behavior of two unsqueezed noise-driven resonators.

7.2.2 Weak Coupling and Strong Pumping

The role of weak coupling becomes more apparent in the presence of strong parametric pumping ($\lambda > \lambda_{\text{th}}$). Each resonator, when pumped above its parametric threshold, is expected to settle into one of its phase states, cf. Fig. 3.3 and Fig. 5.5.[1] The coupling forces between the resonators can cause some phase state configurations to be more attractive than others. We first consider the case that the noise is too small to activate hopping between the phase states, such that the phase state configuration of the system is stable after the initial ringup process.

[1] Note that this simple notion is not generally true, cf. Section 7.3.

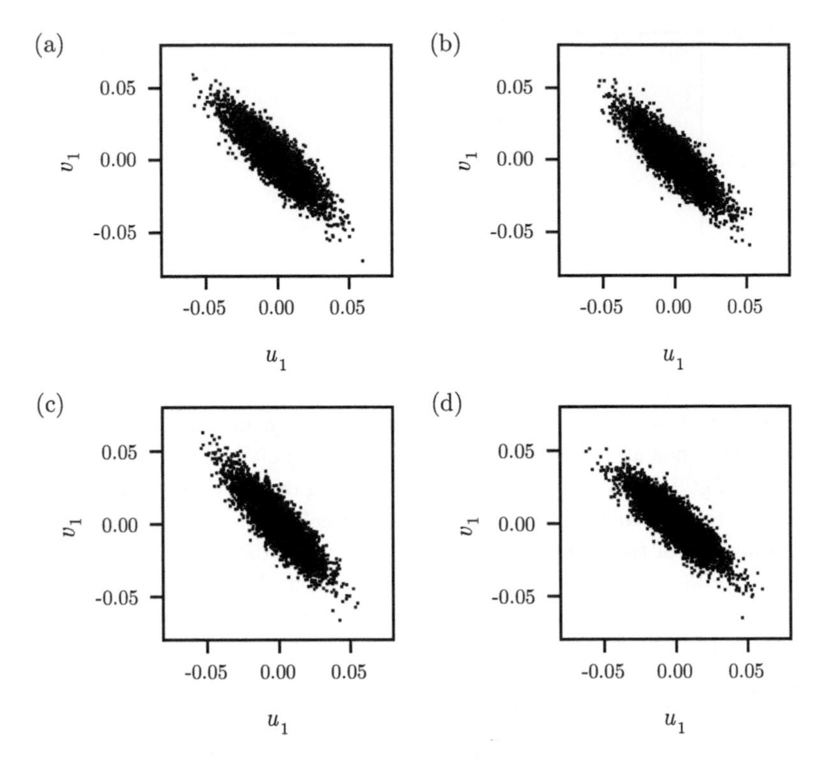

Fig. 7.3 Phase space representation of the squeezed classical fluctuations of two weakly coupled degenerate resonators with $\omega_0 = 100$, $\Gamma = 1$, $m = 1$, $J = 10$, and $\varsigma_D = 1$, that are parametrically driven at $\omega = \omega_0$ with a modulation depth $\lambda = 0.9\lambda_{\text{th}}$, cf. black dot in Fig. 7.2. (a) and (b) show the fluctuations in terms of the original resonator quadratures $u_{1,2}$ and $v_{1,2}$, while (c) and (d) show the fluctuations in a rotated normal mode picture with $u_S = (u_1 + u_2)/\sqrt{2}$, $v_S = (v_1 + v_2)/\sqrt{2}$, $u_A = (u_1 - u_2)/\sqrt{2}$, and $v_A = (v_1 - v_2)/\sqrt{2}$.

For an uncoupled system, each resonator undergoes a separate time-translation symmetry breaking and the relative configuration of the phase states after the ringup is random. Weak coupling, however, biases the system toward a certain *many-body* phase configuration. During the slow ringup process, the two resonators exert a steady force onto each other via the coupling term. The added thermal fluctuations, which can randomize the resonator oscillations, typically change their phase and amplitude on the timescale $\tau_0 = 2\Gamma$. If the ringup process is sufficiently slow, the fluctuations average out (they are effectively low-pass filtered) and even a weak coupling force leads to a deterministic phase configuration of the parametric oscillators.

In Fig. 7.4(a) and (b), we see the response of two weakly coupled resonators ($J > 0$) to a slow parametric sweep from low to high ω. Here, the preferred phase configuration at the time of the ringup process is for the two resonators to be in phase with each other (i.e. the signs of u_1 and u_2 are identical, and likewise for the $v_{1,2}$). We can understand this behavior from the fact that below ω_0, a harmonic resonator responds to an external driving force with a phase of ≈ 0. This phase configuration preference is in agreement

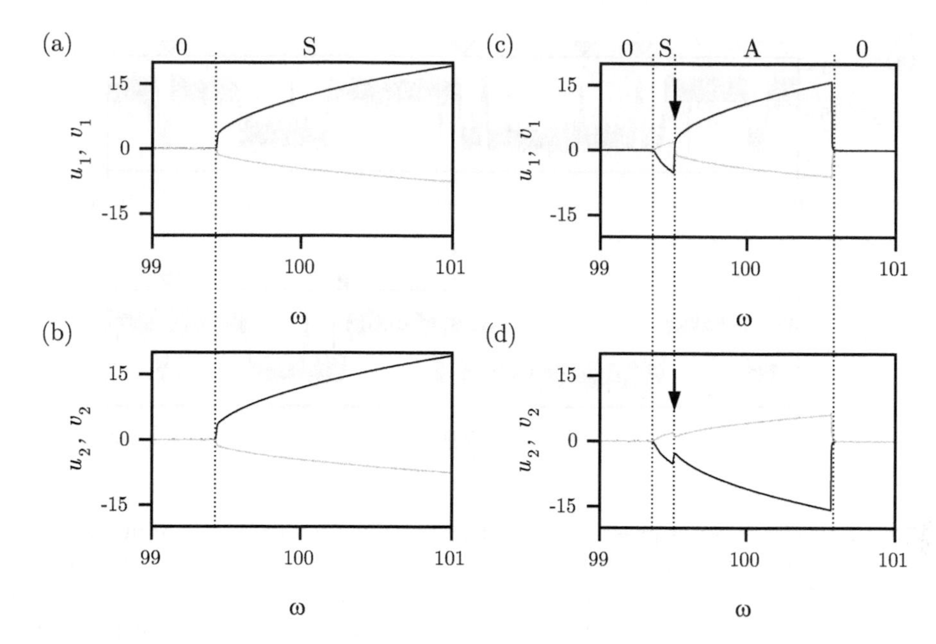

Fig. 7.4 Response of two weakly coupled degenerate resonators to a strong parametric drive. Black and gray lines correspond to $u_{1,2}$ and $v_{1,2}$, respectively, and the system parameters are $\omega_0 = 100$, $\Gamma = 1$, $m = 1$, $\beta = 1$, $\eta = 0$, $J = 10$, $\varsigma_D = 1$, and $\lambda = 1.5\lambda_{\text{th}}$, cf. dotted horizontal line in Fig. 7.2. The total sweep time is 1000. Vertical dotted lines indicate the transitions between regions with symmetric, antisymmetric and zero-amplitude response, labeled S, A and 0, respectively. (a) and (b): sweeping from low to high ω. (c) and (d): sweeping from high to low ω. Arrows mark the transition from an antisymmetric to a symmetric phase state configuration and correspond to the boundary of the antisymmetric (black) Arnold tongue in Fig. 7.2.

with our picture of slightly split normal modes, and with the corresponding Arnold tongues in Fig. 7.2. Once the parametric oscillators have settled into their respective phase states, they persist in that state even beyond the boundary of the Arnold tongue, cf. region III in Fig. 3.2.

When sweeping the parametric pump from high to low ω, the result changes significantly. When the ringup occurs, both resonators are driven at $\omega > \omega_0$, where they respond to a force with a phase lag of $\approx -\pi$. As a consequence, the forces that the resonators apply to each other through the coupling term lead to a preference for the antisymmetric phase configuration, in correspondence with the picture in Fig. 7.2. Here, we sweep ω against the nonlinearity, that is, the parametric oscillator amplitudes become smaller for negative detuning $\omega > \omega_0$. Furthermore, as their phase state amplitudes decrease, the resonators become increasingly sensitive to the influence of their coupling terms and the inverted phase preference. Similar to the case of parametric symmetry breaking due to an external force in Section 3.1.4, the coupling forces start to additionally suppress the phase state amplitudes, leading to a local minimum around $\omega = 99.5$ (see arrows in Fig. 7.4(c) and (d)) and a subsequent switch to a

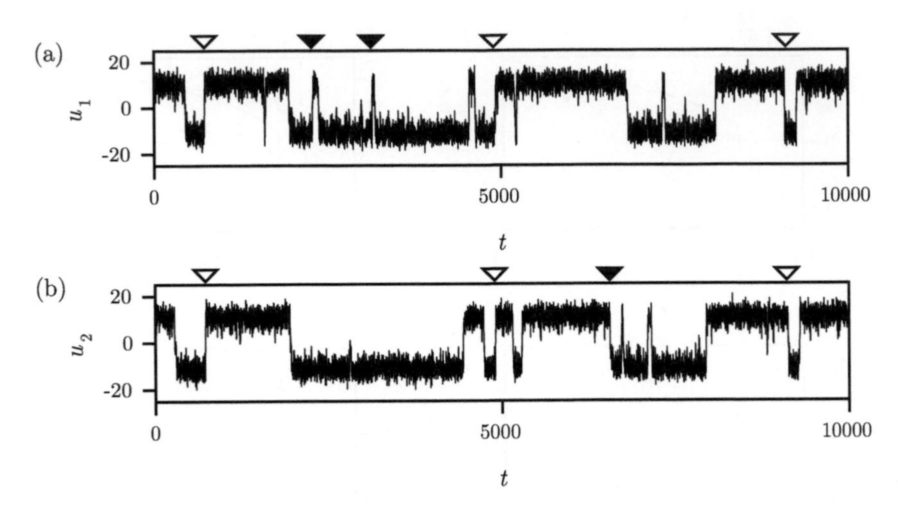

Fig. 7.5 Hopping between phase states under the influence of force noise in a system of two weakly coupled degenerate resonators. Only $u_{1,2}$ are shown in (a) and (b) for simplicity ($v_i \propto -u_i$). The system parameters are $\omega_0 = 100$, $\Gamma = 1$, $m = 1$, $\beta = 1$, $\eta = 0$, $J = 10$, $\varsigma_D = 300$, $\omega = \omega_0$, and $\lambda = 1.5\lambda_{\text{th}}$. White (black) triangles mark examples of hopping events that include both (or only one) of the parametric oscillators.

symmetric phase state configuration. Note that this behavior is exactly what we expect when regarding our coupled system in terms of normal modes. The results shown in Fig. 7.4 are highly reproducible, reflecting the negligible influence of thermal force noise for a slow sweep. We therefore conclude that the normal mode picture, while only rigorously applicable to a model of coupled harmonic oscillators, still retains some of its validity in this regime.

When inverting the sign of J from positive to negative, the phase configurations of the parametric oscillators invert as well, in accordance with the expected normal mode structure. In contrast, the phase configuration becomes random when the sweep is performed very fast (e.g. total time < 100 in our example). This is because the thermal noise is not averaged over a long enough time and therefore gains in relative importance versus the coupling forces.

When the force noise is sufficiently strong, it activates hopping between the phase states of all parametric oscillators. For vanishing coupling between the resonators, the hopping events are uncorrelated, leading to an equal occupation of symmetric and antisymmetric phase configurations over time. In the presence of weak coupling, we expect to observe some bias toward one of the configurations. In Fig. 7.5, we show the simulated behavior of two weakly coupled parametric oscillators with a parametric pump at ω_0, in the center of the overlapping Arnold tongues (also *lobes* in the following). At this position in the stability diagram, both normal modes are stable at similar phase state amplitudes, such that we expect a relatively even statistical distribution between them. In the time traces, we can indeed identify events where both oscillators hop from one phase state to the opposite one, and others where only one oscillator hops.

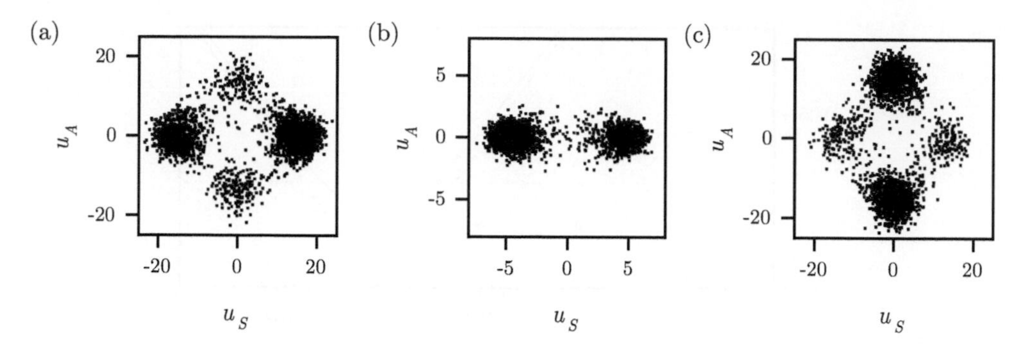

Fig. 7.6 Symmetry phase space representation of two weakly coupled degenerate resonators with large noise. Only $u_{S,A}$ are shown for simplicity ($v_{S,A}$ provide a similar picture). The system parameters are $\omega_0 = 100$, $\Gamma = 1$, $m = 1$, $\beta = 1$, $\eta = 0$, and $\lambda = 1.5\lambda_{\text{th}}$. (a) The coupled system hops between symmetric and antisymmetric configurations for $J = 10$, $\omega = \omega_0$, and $\varsigma_D = 300$, with a preference for the symmetric states. (b) Only symmetric configurations are accessed for $J = 10$, $\omega = 99.44$, and $\varsigma_D = 40$. (c) The antisymmetric states are more favorable for $J = -10$, $\omega = \omega_0$, and $\varsigma_D = 300$.

To visualize the occupation probability of symmetric versus antisymmetric phase configurations, we plot $u_A = (u_1 - u_2)/\sqrt{2}$ as a function of $u_S = (u_1 + u_2)/\sqrt{2}$. Measurement points on the u_S axis signify that the two resonators occupy the same phase state, while points on the u_A axis denote opposite states. The *symmetry phase space* in Fig. 7.6(a) shows a slight preference for a symmetric configuration. This preference becomes more accentuated as the parametric pump is applied at lower frequencies. Eventually, only the symmetric configuration can be accessed, as shown in Fig. 7.6(b). Toward higher frequencies, the symmetric configuration retains a slightly higher occupation. Inverting the sign of J, however, results in a system with a preference for the antisymmetric configuration, see Fig. 7.6(c).

7.2.3 Strong Coupling and Weak Pumping

When the coupling is increased to $\omega_\Delta > \Gamma$, the system enters the strong coupling regime and new phenomena become accessible. Below threshold ($\lambda < \lambda_{\text{th}}$), the lobes are fully separated, see the outlines of the undamped Arnold tongues indicated by dashed lines in Fig. 7.7. We consider small force noise that drives fluctuations of the normal modes, such that nonlinearities can be ignored. Classical squeezing of these fluctuations by a parametric pump can operate on each of the normal modes in the usual way, cf. sections 3.1.1 and 5.2. Note that all phenomena discussed for single resonators apply directly to normal modes as well, once they are sufficiently separated from each other. There is no many-body physics taking place in this regime, and the system dynamics is, in a sense, simpler than that of the corresponding weakly coupled network [76].

In Fig. 7.8, we demonstrate the selective squeezing of normal modes. Inside each lobe, the corresponding normal mode reacts to the parametric pump with squeezed classical fluctuations, while the other normal mode remains unaffected. Pumping the coupled system at $\omega = \omega_0$ has almost no effect on either mode.

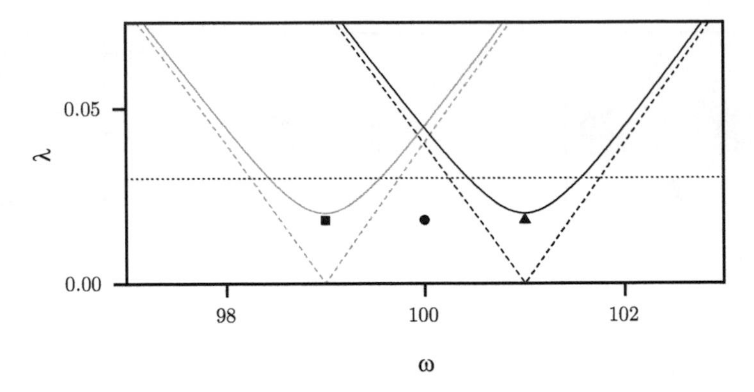

Fig. 7.7 Arnold tongues of a system of two strongly coupled degenerate parametric oscillators. Each resonator has $\omega_0 = 100$, $\Gamma = 1$, and $m = 1$. The static coupling between the resonators is $J = 200$, giving rise to a splitting of $\omega_\Delta = \frac{J}{\omega_0 m} = 2$. Gray (black) lines correspond to the symmetric (antisymmetric) normal modes, while dashed (solid) lines indicate the boundaries of the Arnold tongue without (with) damping, cf. the discussion in Section 3.1.2 and Fig. 3.5. Black symbols indicate the positions of the simulations shown in Fig. 7.8 and a dotted horizontal line marks the frequency sweep shown in Fig. 7.9.

7.2.4 Strong Coupling and Strong Pumping

The notion of normal modes with split Arnold tongues can be carried over to the case of parametric driving above threshold. There exists a range in λ where the lobes do not touch, reaching up to $\lambda = 0.045$ in Fig. 7.7. When performing a sweep of ω in this pumping range, we either recover one or two large-amplitude responses, depending on the direction of the sweep relative to the sign of β, see Fig. 7.9. The upsweep in Fig. 7.9(a) and (b) hits the lower (symmetric) mode first and follows it as long as it remains stable. For our numerical example, we chose $\eta = 0$, which means that the solution would theoretically be stable up to arbitrarily large values of ω. By contrast, the downsweep in Fig. 7.9(c) and (d) allows us to observe both normal modes sequentially, and to visualize their respective symmetries. As we should expect, the upper Arnold tongue corresponds to an antisymmetric configuration of the two resonators, that is, the individual parametric oscillators are in opposite phase states. The difference to the simulations in Fig. 7.4 is the fact that here, the boundaries of the two lobes do not overlap, such that the antisymmetric response is terminated before the symmetric response becomes available.

A partial overlap between the lobes is recovered when the parametric pumping exceeds a second threshold, that is, $\lambda > 0.045$ in Fig. 7.7. Beyond this value, the physics of two strongly coupled parametric oscillators is similar to the weakly coupled case, reflecting the reduced role of the coupling force relative to the parametric pump strength. However, additional effects become prominent in the strong-coupling–strong-pumping regime. These effect are caused by the effective coupling between the normal modes, which is generated by the resonators' nonlinearities (remember that normal modes of harmonic oscillators are not coupled for zero detuning, cf. Section 6.1.1). Rewriting eqn (7.9) in terms of their symmetric (S) and antisymmetric (A) normal

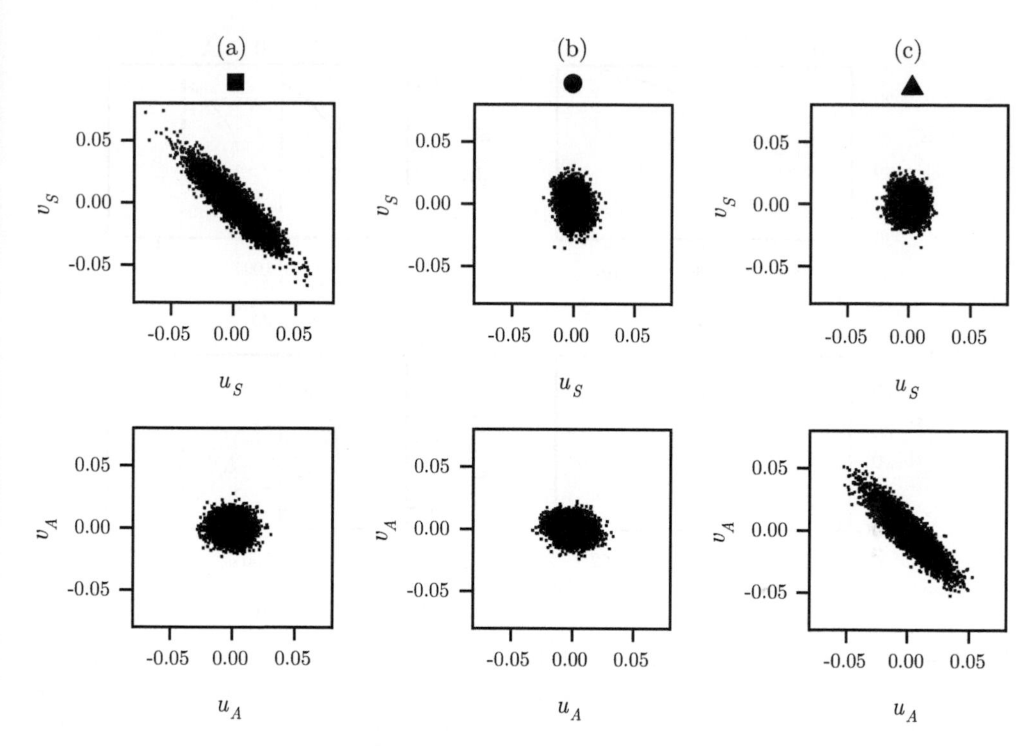

Fig. 7.8 Phase space representation of the symmetric/antisymmetric normal mode fluctuations of two strongly coupled degenerate resonators with $\omega_0 = 100$, $\Gamma = 1$ and $m = 1$, $J = 200$, $\varsigma_D = 1$, and parametrically driven with a modulation depth $\lambda = 0.9\lambda_{\text{th}}$ at (a) $\omega = 99$, (b) $\omega = 100$, and (c) $\omega = 101$, cf. black symbols in Fig. 7.7.

modes, we obtain [86]

$$
\begin{aligned}
\dot{u}_{S,A} =& -\frac{v_{S,A}\left(\frac{3}{4}\beta\left(u_{A,S}^2+v_{A,S}^2\right)+\omega_{S,A}^2-\omega^2\right)}{2\omega}-\frac{3\beta v_{S,A}\left(u_{S,A}^2+v_{S,A}^2\right)}{16\omega} \\
&-\frac{v_{S,A}\left(4\lambda\omega_0^2-3\beta\left(u_{A,S}^2-v_{A,S}^2\right)\right)}{16\omega}-\frac{3\beta u_{A,S}u_{S,A}v_{A,S}}{8\omega}-\frac{\Gamma u_{S,A}}{2}, \\
\dot{v}_{S,A} =& \frac{u_{S,A}\left(\frac{3}{4}\beta\left(u_{A,S}^2+v_{A,S}^2\right)+\omega_{S,A}^2-\omega^2\right)}{2\omega}+\frac{3\beta u_{S,A}\left(u_{S,A}^2+v_{S,A}^2\right)}{16\omega} \\
&-\frac{u_{S,A}\left(4\lambda\omega_0^2-3\beta\left(u_{A,S}^2-v_{A,S}^2\right)\right)}{16\omega}+\frac{3\beta u_{A,S}v_{A,S}v_{S,A}}{8\omega}-\frac{\Gamma v_{S,A}}{2},
\end{aligned}
\tag{7.13}
$$

with $\omega_S^2 = \omega_0^2 - J$ and $\omega_A^2 = \omega_0^2 + J$. We infer from these equations three roles of the nonlinearity β which we will briefly mention here without a deeper discussion. First, the resonance frequencies of the modes are shifted as $\omega_{S,A}^2 \to \omega_{S,A}^2 + \frac{3}{4}\beta\left(u_{A,S}^2 + v_{A,S}^2\right)$. This means that when both modes are driven simultaneously by two independent drive tones, their Arnold tongues shift in frequency in response to each other's amplitudes. Second, we see that the parametric pumping terms are modified according to $4\lambda\omega_0^2 \to 4\lambda\omega_0^2 - 3\beta\left(u_{A,S}^2 - v_{A,S}^2\right)$. When one normal mode reaches a high enough amplitude,

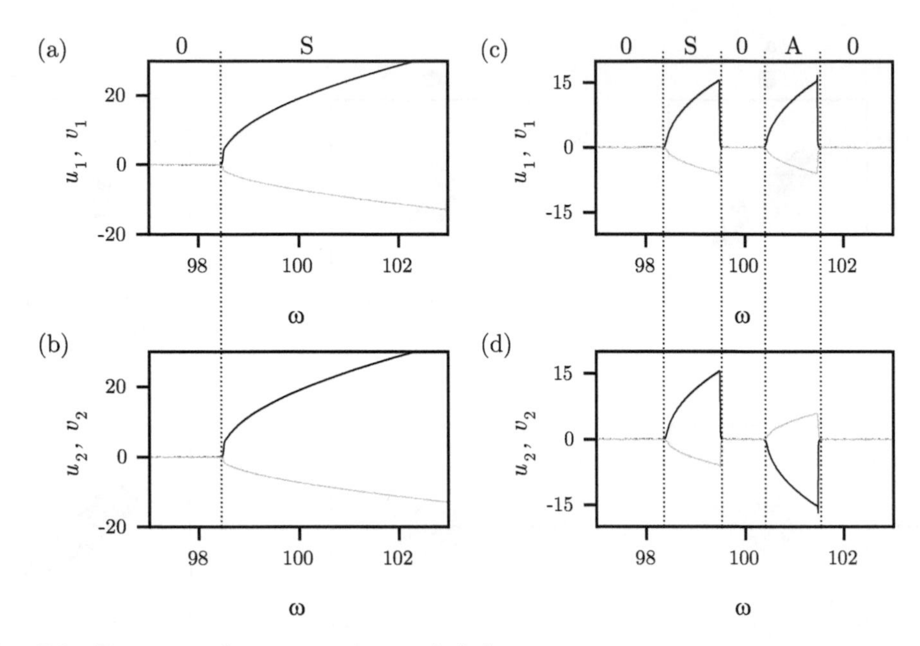

Fig. 7.9 Response of two strongly coupled degenerate resonators to a strong parametric drive. Black and gray lines correspond to $u_{1,2}$ and $v_{1,2}$, respectively, and the system parameters are $\omega_0 = 100$, $\Gamma = 1$, $m = 1$, $\beta = 1$, $\eta = 0$, $J = 200$, $\varsigma_D = 1$, and $\lambda = 1.5\lambda_{\rm th}$, cf. dotted horizontal line in Fig. 7.7. The total sweep time is 1000. Vertical dotted lines indicate the transitions between regions with symmetric, antisymmetric, and zero-amplitude response, labeled S, A, and 0, respectively. (a) and (b): sweeping from low to high ω. (c) and (d): sweeping from high to low ω.

it can generate a parametric instability for the other mode, and an additional system bifurcation appears. As a consequence, *mixed states* comprising both symmetric and antisymmetric oscillations arise in the lobe with higher (lower) frequency for $\beta > 0$ ($\beta < 0$). Third, the nonlinear interaction between the modes also leads to novel bifurcations that are only connected to unstable states. As a result of these *ghost bifurcations*, the mode with the higher (lower) resonance frequency becomes unstable in part of the phase diagram spanned by λ and ω for the case $J > 0$ ($J < 0$) [86]. All of these effects are, in principle, also present in weakly coupled systems, but they are more visible and prominent in the presence of strong coupling.

7.3 Networks with $N > 2$

We finish our phenomenological tour of coupled parametric oscillators with a short outlook toward larger networks.

In a network of identical parametric oscillators with $J \to 0$, all phase states (of the original oscillators 1, 2, and so on) are equivalent in amplitude and differ only by their sign. This is an ideal precondition for using parametric networks as Ising simulators, an idea that has generated significant interest [62, 71, 81, 83, 84, 121]. Such a simulator should allow us to find the ground state of the corresponding Ising Hamiltonian

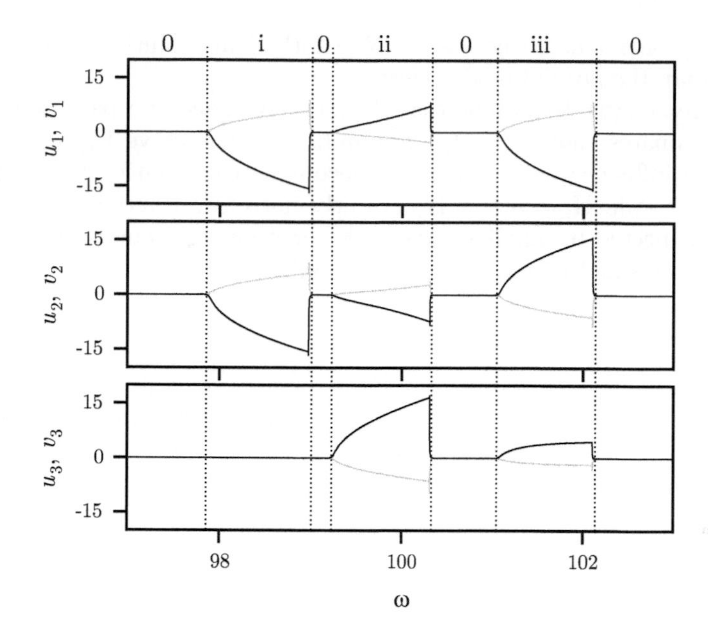

Fig. 7.10 Normal modes (i to iii) of three strongly coupled degenerate parametric oscillators when sweeping ω downward. Black and gray lines correspond to $u_{1,2,3}$ and $v_{1,2,3}$, respectively, and the system parameters are $\omega_0 = 100$, $\Gamma = 1$, $m = 1$, $\beta = 1$, $\eta = 0$, $J_{12} = 600$, $J_{12} = 150$, $J_{12} = -150$, $\varsigma_D = 1$, and $\lambda = 1.5\lambda_{\text{th}}$.

$$H_{\text{Ising}} = -\sum_{i,j} J_{ij}\sigma_i\sigma_j\,, \tag{7.14}$$

where $\sigma_i \in \{-1, 1\}$ is the classical state of a spin (either *up* or *down*) and we use J_{ij} for the coupling between two spins to emphasize the analogy to the resonator network.[2] The two-resonator examples presented so far seem to confirm that a mapping between phase states and spins is possible, because the normal modes of the network with $N = 2$ have the same symmetries as the possible spin configurations; the two phase states of the normal mode labeled S correspond to the spin states up-up and down-down, while the phase states of the mode A correspond to up-down and down-up. However, when moving to $N > 2$, the situation becomes more complicated [86].

The behavior of a network of three parametric oscillators is demonstrated in Fig. 7.10. Three normal modes appear in a frequency sweep, and we can immediately see that the amplitudes of the three resonators are in general not equal. For instance, in the normal mode labeled i, resonator 3 does not oscillate at all. Clearly, this network mode cannot be mapped onto an Ising spin configuration in a simple way. What is more, the number of available phase states is $2N$ (two for each normal mode), while there are 2^N available configurations. For $N = 2$, these two quantities

[2] Note that the physical units of the coupling coefficient between spins is different than in the case of coupled resonators.

are identical by coincidence, whereas for $N > 2$ the Ising solution space grows much more rapidly than the normal-mode space.

It appears reasonable to assume that the solution space of a parametric oscillator network approximates that of an Ising Hamiltonian far above the threshold ($\lambda \gg \lambda_{\mathrm{th}}$), where the influence of the coupling becomes weak relative to the parametric pump [62, 71, 81, 83]. However, it is currently not clear whether this *Ising regime* can always be expected to manifest, and whether it emerges out of the *normal mode regime* close to threshold in a systematic way [86].

Chapter summary

- In Chapter 7, we generalize the coupled systems of Chapter 6 to nonlinear parametric oscillators with force noise.
- We formulate the slow-flow equations of N coupled parametric oscillators, cf. eqns (7.2) to (7.8), and then simplify to the case of two identical, coupled parametric oscillators, cf. eqn (7.9). As long as nonlinearities are weak, the normal-mode picture remains a good description, and we find two Arnold tongues at different frequencies, corresponding to the instability lobes of the symmetric and antisymmetric normal modes, cf. Fig. 7.2.
- It is useful to divide the possible scenarios with respect to two axes: first, we differentiate **weak and strong parametric pumping**, corresponding roughly to squeezing below the parametric threshold and the generation of normal-mode phase states above threshold, respectively. Second, we discriminate **weak and strong coupling**, which is quantified by the normal-mode frequency difference relative to Γ. In most of these cases, we can use our experience from Chapters 3, 5, and 6 to understand the observed phenomena, cf. Figs 7.2 to 7.9. For strong pumping and strong coupling, we briefly mention additional nonlinear effects that can emerge, cf. eqn (7.13) and the discussion thereafter.
- Networks of coupled parametric oscillators are regarded as a potential resource for finding the ground state of **Ising Hamiltonians**. For $N > 2$, an interesting discrepancy arises between the number of normal-mode phase states expected for a network with N oscillators, which is $2N$, and the number of states in an N-spin Ising Hamiltonian, which is 2^N. We illustrate this open question with an example of $N = 3$ parametric oscillators, cf. Fig. 7.10.

Exercises

Check questions:

(a) Consider the case of two identical parametric oscillators pumped below threshold. What is the impact of the different timescales, and especially the relationship between τ_0 and ω_Δ, for the cases with weak and strong coupling?

(b) Can you articulate the role of damping in a parametric oscillator network? What is different for $\Gamma \to 0$?

(c) When hopping occurs between attractors in a parametric oscillator network, how can we qualitatively observe the coupling strength in the time traces? What do you expect for the limit of very strong coupling?

(d) What are the main differences between coupled electron spins and a network of parametric oscillators?

Tasks:

7.1 Open the code **Python Example 7** and study two degenerate, coupled parametric oscillators 1 and 2. First, use $\omega_{1,2} = 100$ ("**w0i = 100*np.array([1,1])**"), $\Gamma_{1,2} = 100$, $\beta_{1,2} = 1$, $\eta_{1,2} = 0$, $\lambda = 1.5\lambda\text{th}$ ("**lam=1.5*2/100**"), $\sigma_m = 1$, and $J = 0.5$ ("**j12=1**" and "**J_coeff=0**") as default values.[3] Perform a slow frequency sweep from **w1=101** to **w1=99** and repeat it several times to obtain some statistics of the resulting phase space distribution. Interpret what you see.

7.2 Repeat frequency sweeps as above for increasing values of the overall coupling **J_coeff** from 0.5 to 50. What are the two consequences of the coupling?

7.3 For various values of **J_coeff**, increase the force noise value σ_m to induce switching between the phase states during the sweep. Balance the force noise such that the phase states are still preserved, that is, the fluctuations do not entirely overwhelm the system. Play with the values of $\beta_{1,2}$ to find different regimes.

7.4 Move to the example of three oscillators. Use the coupling matrix to construct different examples of separated Arnold tongues with distinct symmetries, that is, symmetric, antisymmetric, mixed, and so on. Discuss which of these states can be identified as an approximate Ising state of coupled spins, cf. eqn (7.14).

7.5 Develop a protocol that allows observing all attractors of a network of two parametric oscillators as a function of ω and λ. Use analytical expressions for each Arnold tongue to predict the positions of the bifurcation points, cf. eqn (3.11). Can you find cases where additional bifurcations and solutions appear, as briefly mentioned around eqn (7.13)?

[3] Note that the mass is not explicitly defined and only appears implicitly in the normalized value of the force noise $\sigma_m = \sigma_D/m$. See **Python Example 4** to see how the force noise value can be calculated for a realistic system.

8
The Quantum Harmonic Oscillator

In the previous chapters, we studied deterministic and stochastic classical systems. Now, we widen our scope once more to explore the role of quantum mechanics in parametric phenomena. In preparation for this step, we clarify the required concepts of quantum mechanics on the example of the harmonic oscillator, and will proceed in Chapters 9 and 10 to add dissipation, driving, and nonlinear terms in the quantum language. Our strategy here is to explore why quantum phenomena can differ from classical ones, and to provide a motivation for the methods used to treat them. We adopt an eclectic approach to understanding quantum theory and skip many rigorous derivations in order to focus on our target, that is, the parametric processes presented in Chapter 10. For a comprehensive introduction to quantum mechanics, we refer the reader to the many excellent textbooks on the topic, such as *Quantum Mechanics* by Cohen-Tannoudji, Diu, and Laloe [122].

8.1 From Classical to Quantum Fluctuations

The system we use as a general representation of a harmonic oscillator is a massive particle moving in a parabolic potential, see Fig. 8.1. In Chapter 1, we saw how the time evolution of an isolated particle with an initial displacement x and momentum p can be followed deterministically over its entire future. In Chapters 4 and 5, we found that coupling of an oscillator to an environment leads to classical force noise and to fluctuations of the oscillator variables. Due to the fluctuations, it is impossible to predict where exactly the particle will be at a given time in the future. The uncertainty, however, is only caused by a lack of knowledge concerning microscopic events in the environment, and not by an inherent inability to measure the state of the particle.

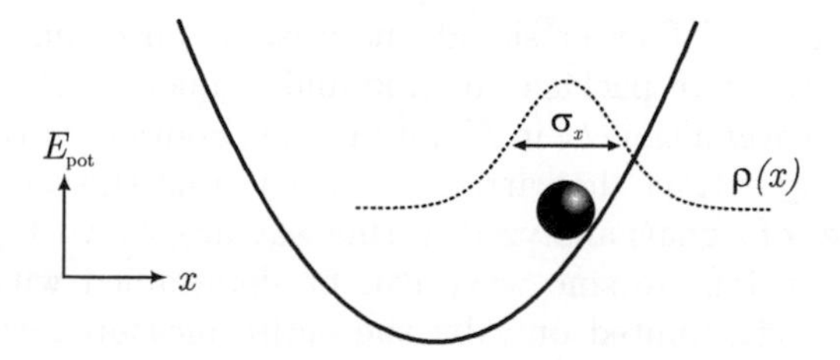

Fig. 8.1 Massive particle in a quadratic potential. Quantum fluctuations introduce a new and inherent uncertainty to the system, illustrated in the example of the displacement x (with a standard deviation σ_x). We can deterministically predict the probability density function $\rho(x)$, which is shown as a dotted line, but not x itself.

8.1.1 The Heisenberg Uncertainty Principle

The role of uncertainty in quantum mechanics is fundamentally different from its classical counterpart. When classical fluctuations are sufficiently small, the behavior of any oscillator is governed by the laws of quantum mechanics. These laws impose a new type of inherent noise to the oscillator state; although a single variable of a quantum oscillator, such as the position x or the momentum p, can be measured with arbitrary precision, the product of the uncertainties of two conjugate (or *complementary*) variables must obey the **Heisenberg uncertainty principle** [122]. For example, the standard deviations of x and p must fulfill

$$\sigma_x \sigma_p \geq \frac{\hbar}{2}. \tag{8.1}$$

A similar relation holds for the energy uncertainty σ_E of a system and its lifetime σ_τ when coupled to an environment,

$$\sigma_E \sigma_\tau \geq \frac{\hbar}{2}, \tag{8.2}$$

which is known as the **Mandelshtam–Tamm inequality** [123]. The meaning of the term lifetime is related to the process of energy decay in a classical resonator, cf. Chapter 1. There, energy is lost from an oscillating massive particle at a rate $\Gamma = Q/\omega_0$ and with an exponential ringdown $x^2(t) \sim e^{-t\Gamma}$, cf. eqn (1.6). For the equivalent quantum system, it turns out that energy decay to an environment can be described by the same exponential form, and that $\sigma_\tau = \Gamma^{-1}$ [124]. The lifetime is therefore tied to the rate at which a system decays or relaxes, which we will revisit in Section 9.2.2.

These uncertainty principles do not follow from the details of specific experiments, but represent an intrinsic quality of nature. They require the **reduced Planck constant** $\hbar \approx 1.055 \times 10^{-34}$ Js, which was originally introduced to determine the energy of each photon in a light wave of frequency ω, namely $E_{\text{photon}} = \hbar\omega$. This famous formula, which historically stood at the very beginning of the discovery of quantum mechanics, will naturally arise from the Heisenberg uncertainty principle in the following sections for the case of a harmonic oscillator.

One way to approach the uncertainty principles from the classical world is to compare them with classical Fourier signal analysis. As an example, let us consider the width of an acoustic wave package in time and frequency. One can measure the wave displacement $x(t)$ over a time t_S in N data points, Fourier transform the resulting sample path, and finally extract the carrier frequency from the peak in the spectrum of $x(\omega)$. Two examples of signals analyzed in this way are shown Fig. 8.2.

The frequency of an infinite sine wave can be determined with arbitrary precision with Fourier analysis, limited only by the entire measurement duration t_S, cf. eqn (4.15). Any amplitude modulation of the wave, however, leads to additional spectral components in the Fourier transform [125]. The peaks in Fig. 8.2(b) are therefore broadened by the continuous amplitude modulations of the carrier frequency in Fig. 8.2(a). A short pulse in time (black trace) leads to a broad Fourier spectrum, while a longer pulse (gray trace) retains a narrower distribution in frequency. The

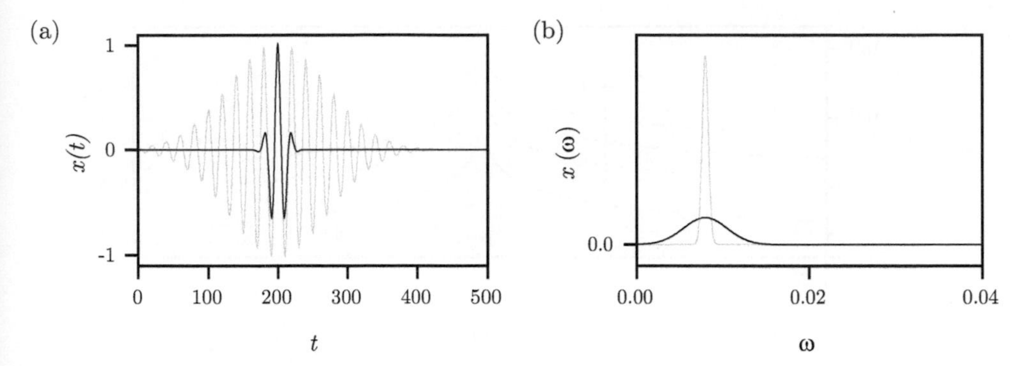

Fig. 8.2 (a) Time traces of two signals shown in gray and black. (b) Fourier transforms of the signals in (a) with arbitrary vertical scaling.

broadening in frequency and time can be defined via the standard deviations σ_ω and σ_τ of the package extension in both dimensions. For instance, the damped harmonic oscillator with a thermal drive that we studied in Chapter 4 corresponds to an overlapping series of wave packages with $\sigma_\tau = \Gamma^{-1}$ and $\sigma_\omega = \Gamma$, and thus to $\sigma_\tau\sigma_\omega = 1$. There is, however, a lower bound for any signal known as **Gabor's limit** corresponding to $\sigma_\tau\sigma_\omega \geq \frac{1}{2}$ [126, 127]. For brevity, we do not present the derivation of this tighter bound. Note that Gabor's limit leads exactly to eqn (8.2) when inserting a Planck–Einstein relation for the broadening, $\sigma_E = \hbar\sigma_\omega$, and interpreting the package broadening as an uncertainty. We therefore only require classical signal analysis and knowledge of the quantization (or *graininess*) of waves to understand the uncertainty relation between energy and time. An analogous Fourier analysis can be carried out for the position x of a wave and its momentum p, leading ultimately to eqn (8.1).

Please note that quantum mechanics does not forbid measurement of either x or p with arbitrary precision. However, it is not possible to measure x without disturbing p, and vice versa. This perturbation causes the conjugate variable to be *projected* into a state with a sufficiently large uncertainty to fulfill the Heisenberg uncertainty principle at all times [122]. Such projective (or *strong*) measurements are crucial for quantum sensing, but will not be examined in depth in this text, see instead Refs [128, 129].

8.1.2 The Ground State Energy

In Chapter 4, we discussed how classical fluctuating forces are generated when an oscillator is coupled to an environment with temperature T. The fluctuating forces lead to uncertainties in x and p and to an average equilibrium energy of

$$\langle E_{\text{eq}} \rangle = m\omega_0^2 \langle x^2 \rangle = k_B T\,. \tag{8.3}$$

In analogy with this picture, we can associate the Heisenberg uncertainty principle with **quantum fluctuations**. Due to such fluctuations, a harmonic quantum oscillator that is devoid of any classical force noise ($T \to 0$) will still fluctuate with an average energy of

$$\langle E_0 \rangle = m\omega_0^2\sigma_x^2 = \omega_0\sigma_x\sigma_p \geq \frac{1}{2}\hbar\omega_0 \tag{8.4}$$

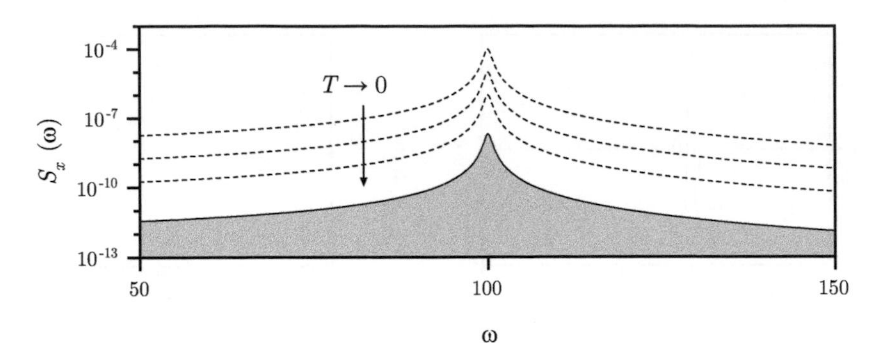

Fig. 8.3 Illustration of the displacement fluctuation PSD S_x of a harmonic oscillator. Dashed lines represent the thermal PSD for decreasing temperature T (from top to bottom). For $T \to 0$, the displacement fluctuations of the oscillator converge toward their lower limit given by the ground state energy, shown as a solid line. The area under this line amounts to $E_0 = \hbar\omega_0/2$.

where we have made use of the relation $\sigma_p = \sigma_x m\omega_0$, and of the definitions $\sigma_x^2 = \langle x^2 \rangle - \langle x \rangle^2$ and $\sigma_p^2 = \langle p^2 \rangle - \langle p \rangle^2$ for $\langle x \rangle = \langle p \rangle = 0$. This minimal resonator energy E_0 is the so-called **ground state energy** (or zero-point energy) of a quantum resonator that can be measured even for macroscopic objects [130]. The Heisenberg uncertainty principle is thus not merely a limitation of what we can know about a state, but it leads to concrete quantum fluctuations with a well-defined average energy, similar to the thermal fluctuations in the classical case, see Fig. 8.3 for a schematic illustration.

8.2 From First to Second Quantization

The uncertainty imposed by quantum fluctuations makes it necessary to describe the oscillator state with the help of a probability density function $\rho(x,t)$, just as we did in Section 4.3 for a stochastic classical resonator. Previously, the uncertainty was a manifestation of coupling to an environment which imparted stochastic forces onto the system via the fluctuation–dissipation theorem, cf. eqn (4.25). In this sense, $\rho(x,t)$ described the probability of observing the resonator at a specific coordinate in phase space when repeating the experiment several times. In the presence of quantum fluctuations, $\rho(x,t)$ describes the same information, but the source of uncertainty is inherent and not imposed by the environment. Due to this inherent uncertainty, we interpret the probability density not as a lack of information, but as a so-called **superposition** of different values, that is, the resonator has the potential to simultaneously reside in different states.

8.2.1 The Schrödinger Equation

As just stated, the intrinsic nature of quantum fluctuation implies that a quantum system is best described by a probability density function $\rho(x,t)$. Interestingly, we know from wave mechanics that such a probability density field can also be written as a wave's power density $\rho(x,t) = |\Psi(x,t)|^2$, where $\Psi(x,t)$ is a wave function. In classical wave physics, the wave function describes the collective amplitude of a gas cloud of

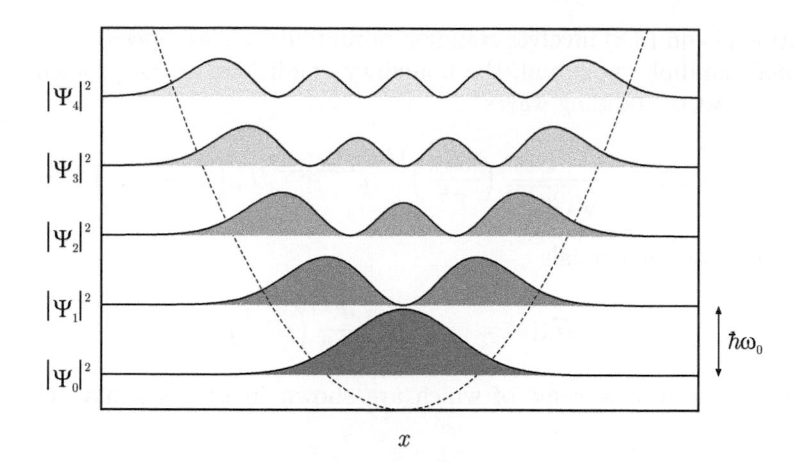

Fig. 8.4 Representation of the squared wave functions of the first five Fock states (shaded areas) in a parabolic potential well (dashed line). The baseline of each shaded function corresponds to the eigenenergy of the corresponding Fock state. The baselines are spaced by $\hbar\omega_0$.

particles. The wave function is therefore identified as a *probability amplitude* that can be used to calculate the chance of finding our system in a certain state. Obviously, as the wave function describes a single particle, $\Psi(x,t)$ is normalized to fulfill

$$\int_{-\infty}^{\infty} |\Psi(x,t)|^2 \, dx = 1\,. \tag{8.5}$$

We have reached a wave description of a single particle. This **particle-wave duality** is one of the foundational ideas of quantum mechanics [131]. Importantly, while the evolution of a particle's position x or momentum p can only be predicted as a probability, the wave function Ψ itself can be calculated in a deterministic fashion. This key concept is captured in the deterministic differential equation known as the **Schrödinger equation**

$$i\hbar\frac{\partial}{\partial t}\Psi(x,t) = H\Psi(x,t)\,, \tag{8.6}$$

with H being the system's Hamiltonian. For example, for a massive particle in a parabolic potential well, the Hamiltonian reads

$$H = -\frac{\hbar^2}{2m}\frac{\partial^2}{\partial x^2} + \frac{1}{2}m\omega_0^2 x^2\,. \tag{8.7}$$

The Schrödinger equation is a partial differential equation describing the evolution in space and time of a wave function $\Psi(x,t)$. Assuming a linear Hamiltonian, we can employ standard separation of variables, and the time-independent part of the Schrödinger equation reads

$$\hat{H}\psi(x) = E\psi(x)\,. \tag{8.8}$$

The solution of eqn (8.8) involves standing probability waves. Specifically, for our one-dimensional parabolic potential, the boundary conditions of the problem allow only for a discrete set of standing waves

$$\Psi_n(x) = \frac{1}{\sqrt{(2^n n!)}} \left(\frac{m\omega_0}{\pi\hbar}\right)^{1/4} e^{-\frac{m\omega_0 x^2}{2\hbar}} H_n\left(\sqrt{\frac{m\omega_0}{\hbar}}x\right), \tag{8.9}$$

with the Hermite polynomials

$$H_n(z) = (-1)^2 e^{z^2} \frac{d^n}{dz^n}\left(e^{-z^2}\right). \tag{8.10}$$

These discrete solutions, some of which are shown in Fig. 8.4, have evenly-spaced eigenenergies

$$E_n = \hbar\omega_0 \left(\frac{1}{2} + n\right). \tag{8.11}$$

This is the discretized energy spectrum of the **quantum harmonic oscillator**. It is an important milepost on our path because it shows that resonators in the quantum regime have a discrete set of energy eigenvalues and eigenstates, which are often referred to as **Fock states**. Equation (8.11) also confirms, for the particular case of a standing wave, the Planck–Einstein relation $E_{n+1} - E_n = \hbar\omega_0$ if we postulate that jumps between discrete Fock states correspond to the addition or subtraction of one energy quantum to our resonator. The energy quanta are identified as the corresponding particle manifestations of the wave, for example photons or phonons. This notion of discrete energy states and particles will guide us to the idea of a vector notation and the second quantization in the following sections.

At this point, we pause to make a clear distinction between concepts that may easily be confused. Let us consider a string clamped between two points as the physical realization of our harmonic oscillator. Along its y-axis, the string can form standing waves that define effective modes at frequencies ω_i. We would like to emphasize that these different modes are *not* associated with different Fock states, even though the Fock states $\Psi_n(x)$ and the mode shapes can look very similar. Rather, each of the string modes corresponds to a separate Hamiltonian of the form

$$\hat{H}_i = -\frac{\hbar^2}{2m}\frac{\partial^2}{\partial x^2} + \frac{1}{2}m\omega_i^2 x^2, \tag{8.12}$$

each with eigenvalues

$$E_{i,n_i} = \hbar\omega_i \left(\frac{1}{2} + n_i\right). \tag{8.13}$$

The energy of every individual mode is described by the Fock index n_i, which marks the nth rung of the Fock ladder in a potential well with frequency ω_i. An illustration of this situation is provided in Fig. 8.5. When we deal with a single oscillator, we drop the index i and simply use n for the Fock states.

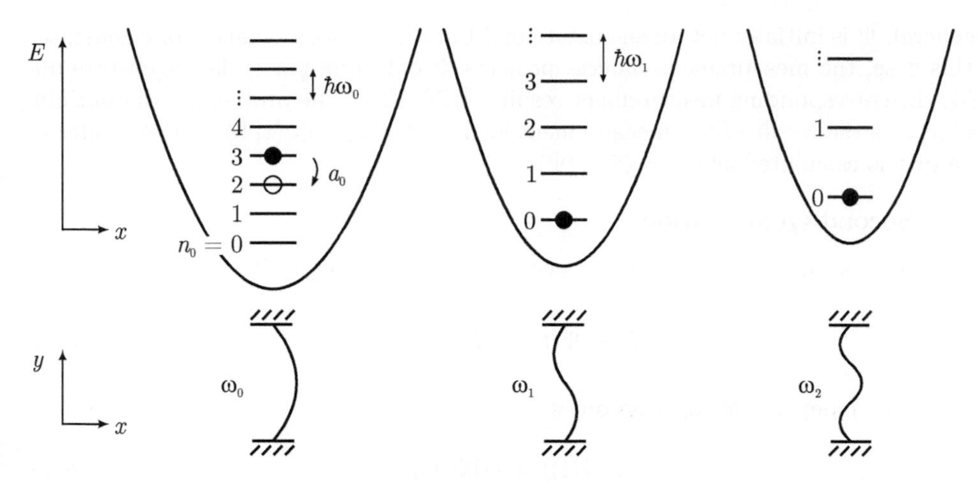

Fig. 8.5 Visualization of three oscillation modes of a string. Each mode has a discrete set of energy levels that correspond to eigenstates of the Hamiltonian of the mode (Fock states with their respective n_i on the left). In the example shown here, mode 0 is in the Fock state with $n_0 = 3$ and is about to be transported to $n_0 = 2$ by the annihilation operator $\hat{a}_0$ (see Section 8.2.3). The two higher modes are both in their lowest Fock state $n_{1,2} = 0$, which corresponds to the vacuum ground state.

8.2.2 Dirac's Notation

Having seen the inherent quantization (discretization) of quantum states, we can translate the wave function into a vector (Hilbert) space whose elements are formed by all possible values of the parameters of Ψ. In Dirac's *bra–ket* notation, a vector z in this space is written as *ket* $|z\rangle$. For example, an infinite number of position vectors $|x\rangle$ identify all possible x-coordinates in a given system. The x-dependent wave function $\Psi(x)$ at a particular time can then be expressed as the state vector

$$|\Psi\rangle = \int \Psi(x)\,|x\rangle\,dx\,. \tag{8.14}$$

The ket $|\Psi\rangle$ is thus an infinite vector that lists the projections of $\Psi(x)$ onto some basis $|j\rangle$, for instance the position basis $|x\rangle$. We can read the amplitude of $\Psi(x)$ at a position x_0 directly from the corresponding entry of $|\Psi\rangle$, which can be extracted by a multiplication with the *bra* vector $\langle x_0| = |x_0\rangle^\dagger$ (where † denotes the Hermitian conjugate operation),

$$\Psi(x_0) = \langle x_0|\Psi\rangle\,. \tag{8.15}$$

In the wave mechanics formalism of quantum mechanics that we use here, a measurable quantity z is therefore associated with an operator $\hat{z}$. Applying an operator to a wave function Ψ represents a measurement process. If Ψ is already an eigenstate of $\hat{z}$, the measurement simply yields the corresponding eigenvalue z, which fulfills

$$\hat{z}\,|\Psi\rangle = z\,|\Psi\rangle\,. \tag{8.16}$$

In general, Ψ is initially not an eigenstate of $\hat{z}$ but a linear combination of eigenstates. In this case, the measurement process projects Ψ onto one particular eigenstate and yields the corresponding measurement result z [128, 129]. The probability of obtaining a value z as the result of the measurement is given by $\Psi(z) = \langle z|\Psi\rangle$. The expectation value of z is calculated as $\langle \hat{z}\rangle = \langle\Psi|\,\hat{z}\,|\Psi\rangle$.

8.2.3 Second Quantization

In the notation of Dirac, the time-dependent Schrödinger equation is

$$i\hbar\frac{\partial}{\partial t}\left|\Psi(t)\right\rangle = \hat{H}\left|\Psi(t)\right\rangle \tag{8.17}$$

and its time-independent form becomes

$$\hat{H}\left|\Psi(t)\right\rangle = E\left|\Psi(t)\right\rangle \,, \tag{8.18}$$

where $\hat{H}$ is now an operator that acts on the quantum wavefunction. Remember that the solutions of eqn (8.18) are the discrete energy (or Fock) states. In the following, we will use the notation $|n\rangle$ to indicate the nth Fock state, and $|0\rangle$ corresponds to the ground state with energy $\frac{1}{2}\hbar\omega_0$. The states $|n\rangle$ form the orthonormal **Fock basis.**

We can define additional useful operators that act on our quantum state. In the Fock representation, for example, transitions between adjacent levels can be assigned to an annihilation operator $\hat{a}$ (which lowers n by 1) and a creation operator $\hat{a}^\dagger$ (which increases n by 1),

$$\begin{aligned}\hat{a}\left|n\right\rangle &= \sqrt{n}\left|n-1\right\rangle \,,\\ \hat{a}^\dagger\left|n\right\rangle &= \sqrt{n+1}\left|n+1\right\rangle \,.\end{aligned} \tag{8.19}$$

Application of both of these so-called *ladder operators* (one after the other) keeps the particle number unchanged and defines the *number operator* $\hat{n} \equiv \hat{a}^\dagger\hat{a}$ whose eigenvalue is n,

$$\hat{a}^\dagger\hat{a}\left|n\right\rangle = \hat{n}\left|n\right\rangle = n\left|n\right\rangle \,. \tag{8.20}$$

Through straightforward calculations [122], additional operators can be defined, where the following relations apply

$$\begin{aligned}\hat{a} &= \sqrt{\frac{m\omega_0}{2\hbar}}\left(x + \frac{i}{m\omega_0}p\right) \,,\\ \hat{a}^\dagger &= \sqrt{\frac{m\omega_0}{2\hbar}}\left(x - \frac{i}{m\omega_0}p\right) \,,\\ \hat{x} &= \sqrt{\frac{\hbar}{2m\omega_0}}\left(\hat{a}^\dagger + \hat{a}\right) \,,\\ \hat{p} &= i\sqrt{\frac{\hbar m\omega_0}{2}}\left(\hat{a}^\dagger - \hat{a}\right) \,.\end{aligned} \tag{8.21}$$

Using these relations, we can now write our Hamiltonian in eqn (8.7) in terms of ladder operators.[1] This procedure is called **second quantization**. Using eqns (8.21), our Hamiltonian from eqn (8.7) becomes

$$H = \hbar\omega_0 \left(a^\dagger a + \frac{1}{2}\right), \tag{8.22}$$

which has the eigenenergies E_n given in eqn (8.11). This is a result of the fact that the Hamiltonian describes an undriven and undamped resonator, such that the particle number in the system remains conserved. As such, the Fock state basis is an eigenbasis of eqn (8.22).

As we will discuss in Chapter 9, in open systems the particle number is in general not conserved and we will rely on another basis to describe our system. In preparation for the open system case, let us start by considering the eigenstate of the ladder operator a, dubbed the **coherent state** $|a\rangle$. It has a complex eigenvalue that we denote by α, and it possesses an average number of excitations: measuring the number of excitations in a coherent state, we find $\langle n\rangle = \langle a^\dagger a\rangle = |\alpha|^2$. Furthermore, a coherent state has equal uncertainty in its dimensionless x and p operators

$$x_{\mathrm{dl}} \equiv \sqrt{\frac{m\omega_0}{2\hbar}}x = \frac{1}{2}\left(a^\dagger + a\right), \tag{8.23}$$

$$p_{\mathrm{dl}} \equiv \sqrt{\frac{1}{2m\hbar\omega_0}}p = \frac{i}{2}\left(a^\dagger - a\right), \tag{8.24}$$

and therefore a minimum total (Heisenberg) uncertainty $\sigma_x\sigma_p$. It can be regarded as a Fock state $|0\rangle$ that is displaced from the center of phase space, or as a quantum superposition of all Fock states following the Poisson distribution

$$|\alpha\rangle = e^{-\frac{|\alpha|^2}{2}} \sum_{n=0}^{\infty} \frac{\alpha^n}{\sqrt{n!}} |n\rangle . \tag{8.25}$$

By these definitions, $\alpha = x_{\mathrm{dl}} + ip_{\mathrm{dl}}$ is the complex amplitude of the coherent state, and the dimensionless operators x_{dl} and p_{dl} are its quadratures. We can compare the squared amplitude with that from eqn (4.33), $|\alpha|^2 = X^2$, and obtain with eqns (8.21)

$$X^2 = \frac{\hbar}{2m\omega_0}\left((a^\dagger + a)^2 - (a^\dagger - a)^2\right) = \frac{2\hbar}{m\omega_0}(n - \frac{1}{2}). \tag{8.26}$$

In the classical limit of large n, $\langle X^2\rangle$ and $\langle n\rangle$ are thus proportional. Inserting eqn (8.26) into the Hamiltonian in eqn (8.22) for $n = a^\dagger a \gg 1$, we recover the correct total energy

$$E = \frac{1}{2}m\omega_0^2 X^2. \tag{8.27}$$

[1] To simplify the notation, we drop the hat symbol and generally write a instead of $\hat{a}$ etc. in the following.

8.2.4 Mixed States and the Master Equation

So far, our discussion has revolved around so-called *pure* states that can be described by a single wave function $\Psi(x,t)$. In many realistic situations, this may no longer be possible, for instance because it is not known accurately in which state the system was prepared. In such cases, several wave functions are needed for a full representation and the probability density is obtained from their weighted statistical mixture,

$$\rho(x,t) = \sum_i P_i \left|\Psi_i(x,t)\right|^2 , \tag{8.28}$$

where the P_i, whose sum must be 1, indicate the relative probabilities that the system is in the corresponding pure state $\Psi_i(x,t)$. Equation (8.28) describes a so-called **mixed state** when there are at least two $P_i \neq 0$. In order to deal with mixed states in Dirac's vector notation, we introduce the **density matrix** (or density operator)

$$\rho = \sum_i P_i \left|\Psi_i\right\rangle \left\langle\Psi_i\right| . \tag{8.29}$$

The Schrödinger equation does not apply to mixed states. How can we describe the time evolution of ρ instead? For the example of a simple density matrix $\rho = \left|\Psi\right\rangle\left\langle\Psi\right|$, we find

$$\frac{\partial\rho}{\partial t} = \frac{\partial}{\partial t}\left(\left|\Psi\right\rangle\left\langle\Psi\right|\right) = \left(\frac{\partial}{\partial t}\left|\Psi\right\rangle\right)\left\langle\Psi\right| + \left|\Psi\right\rangle\left(\frac{\partial}{\partial t}\left\langle\Psi\right|\right) . \tag{8.30}$$

Inserting eqn (8.17) with $i\hbar\frac{\partial}{\partial t}\left\langle\Psi\right| = -H\left\langle\Psi\right|$ allows us to write eqn (8.30) as

$$\frac{\partial\rho}{\partial t} = \frac{1}{i\hbar}H\left|\Psi\right\rangle\left\langle\Psi\right| - \frac{1}{i\hbar}\left|\Psi\right\rangle\left\langle\Psi\right|H , \tag{8.31}$$

and we arrive at the form

$$i\hbar\frac{\partial\rho}{\partial t} = [H,\rho] \tag{8.32}$$

with the commutator operator defined as

$$[A,B] = AB - BA . \tag{8.33}$$

The entire procedure also works with density matrices composed out of a sum of pure states. Equation (8.32) is the **von Neumann master equation** that allows us to follow the evolution of entire density matrices. It is very similar to Liouville's theorem in eqn (4.37), with the difference that von Neumann's equation replaces the Poisson bracket by the operator commutator, cf. eqn (8.33).

The expectation value of an operator z, which we identified as $\langle z\rangle = \left\langle\Psi\right| z \left|\Psi\right\rangle$ for a pure state, is now calculated as

$$\langle z\rangle = \mathrm{tr}(\rho z) . \tag{8.34}$$

8.3 Quantum State Representations

8.3.1 Definitions

We have seen how to evolve a (closed) quantum mechanical system in time, cf. eqns (8.17) and (8.32). Similar to the classical EOM, we can numerically evolve the quantum EOM for all elements of the system's quantum state or density matrix. Recall, however, that the Hilbert space is of infinite dimension even for a single harmonic resonator. To make a numerical treatment possible, we commonly truncate the dimension of the Hilbert space and make sure that our numerical simulation remains within the reduced space.

Below, we will show several illustrative examples of how a quantum harmonic resonator evolves in time given different initial boundary conditions. Nowadays, there are several numerical packages that implement such numerical quantum time evolution algorithms, and the figures in this chapter were produced with the help of QuTiP [132], using a Fock basis for a density matrix that is truncated at high photon numbers. To visualize our quantum states as they evolve, we employ a number of projected observables; in many cases, we will present results in terms of the dimensionless position and momentum operators x_{dl} and p_{dl}, cf. eqn (8.23). The observables are given by the traces $\langle x_{\mathrm{dl}}\rangle = \mathrm{tr}\{\rho\, x_{\mathrm{dl}}\}$ and $\langle p_{\mathrm{dl}}\rangle = \mathrm{tr}\{\rho\, p_{\mathrm{dl}}\}$, respectively. The inherent quantum uncertainty of observing x_{dl} and p_{dl} at a given time is quantified by their associated standard deviation operators,

$$
\begin{aligned}
\sigma_{x_{\mathrm{dl}}} &= \sqrt{\mathrm{tr}\{\rho\,(x_{\mathrm{dl}} - \langle x_{\mathrm{dl}}\rangle)^2\}}\,, \\
\sigma_{p_{\mathrm{dl}}} &= \sqrt{\mathrm{tr}\{\rho\,(p_{\mathrm{dl}} - \langle p_{\mathrm{dl}}\rangle)^2\}}\,,
\end{aligned}
\tag{8.35}
$$

respectively. Furthermore, we study the outcome of the particle-number operator n over time.

In previous chapters, we have seen the merit of looking at the resonator and its probability distribution function in phase space. There are corresponding quantum operators such as the **Husimi-Q distribution**, which for a phase space spanned by x and p is defined as

$$
Q_H(\alpha) = \frac{1}{\pi}\,\langle\alpha|\,\rho\,|\alpha\rangle\,,
\tag{8.36}
$$

meaning that it probes how much the density matrix ρ resembles a coherent state $|\alpha\rangle$. This provides a quasi-probability density over the over-complete coherent state basis. Plotting the Husimi-Q distribution over the phase-space coordinates matches the Fokker–Planck distribution function in the semiclassical limit.

The phase space representation that we will predominantly use is the **Wigner quasiprobability distribution** [133, 134], which is defined as

$$
W(x,p) = \frac{1}{\pi\hbar}\int_{-\infty}^{\infty} \langle x+y|\,\rho\,|x-y\rangle\, e^{-i2py/\hbar} dy\,.
\tag{8.37}
$$

For a pure state, $W(x,p)$ is bounded by the values $-\frac{2}{\hbar} < W(x,p) < \frac{2}{\hbar}$ as a consequence of the Heisenberg uncertainty principle. In a classical setting with $\hbar \to 0$, this limit

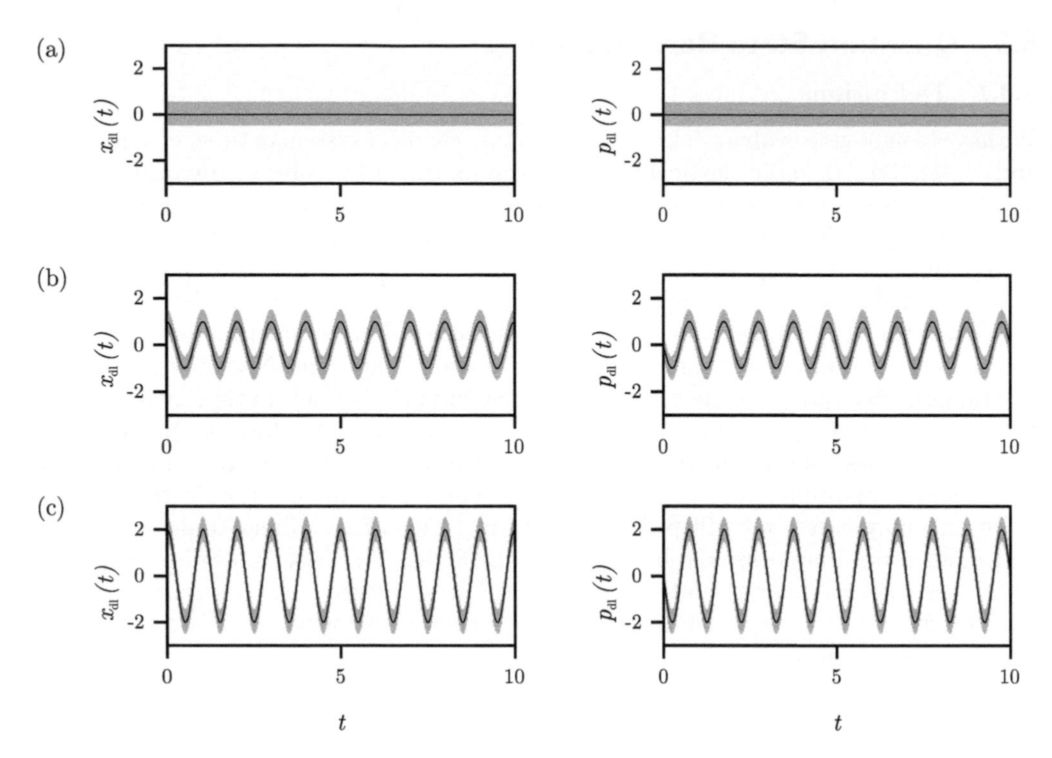

Fig. 8.6 Time evolution using a bare harmonic oscillator, cf. eqn (8.32). The expectation values of x and p as a function of time are plotted as black lines for an initial density matrix of the Fock state (a) $|\alpha| = 0$, (b) $|\alpha| = 1$, and (c) $|\alpha| = 2$. Green shaded areas indicate the corresponding standard deviations, cf. eqns (8.35).

drops and the Wigner function can be concentrated in an arbitrarily small part of phase space and with arbitrarily large absolute values.

In terms of the dimensionless amplitude quadratures x_{dl} and p_{dl}, the Wigner quasiprobability distribution can be calculated according to

$$W(x_{\mathrm{dl}}, p_{\mathrm{dl}}) = \frac{1}{\pi} \int_{-\infty}^{\infty} \langle x_{\mathrm{dl}} + y | \, \rho \, | x_{\mathrm{dl}} - y \rangle \, e^{-i2p_{\mathrm{dl}}y} dy \, . \tag{8.38}$$

Note the missing factor $\hbar$ in the normalization compared to eqn (8.37) due to the fact that x_{dl} and p_{dl} are unitless.

The Wigner distribution is popular as an indicator for non-classical parts in a density matrix, which appear as negative values. Interestingly, it was shown [135] that positive and negative parts of $W(x,p)$ are related to the symmetric (Ψ_S) and antisymmetric (Ψ_A) contributions around this coordinate of a wave function $\Psi = \Psi_S + \Psi_A$. Hence, the Wigner distribution can be rewritten as

$$W(x,p) = \frac{1}{\pi\hbar} \left(|\Psi_S|^2 - |\Psi_A|^2 \right) . \tag{8.39}$$

Equation (8.39) facilitates an interpretation of the negative values in the Wigner distribution. Maximum negativity should be expected when $\Psi(x,p)$ is locally antisymmetric,

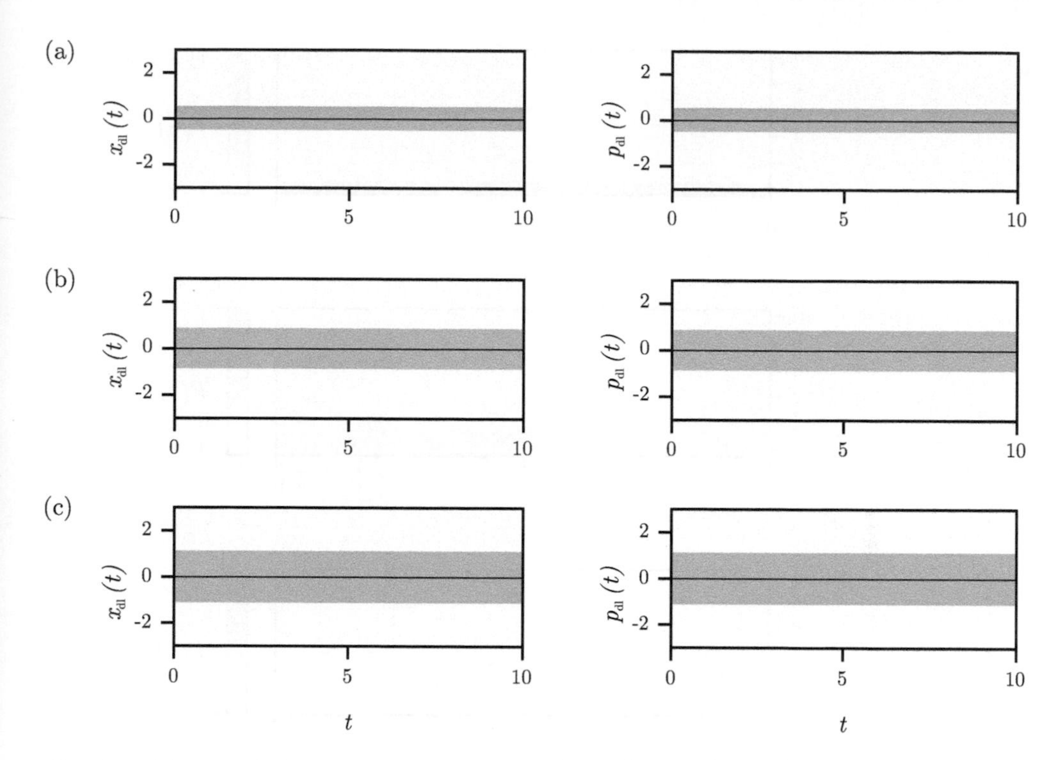

Fig. 8.7 Time evolution using a bare harmonic oscillator, cf. eqn (8.32). The expectation values of x and p as a function of time are plotted as black lines for an initial density matrix of the Fock state (a) $|n = 0\rangle$, (b) $|n = 1\rangle$, and (c) $|n = 2\rangle$. Green shaded areas indicate the corresponding standard deviations, cf. eqns (8.35).

that is, where it crosses through $\Psi(x, p) = 0$. It is worth noting that the Husimi-Q representation (8.36), which contains only positive values, can be regarded as a Wigner distribution function that is smoothed out over the Gaussian peak of its coherent probe state. This smoothing precludes the detection of negative values, which can only exist on scales smaller than the peak width.

8.3.2 Visualization of Quantum State Evolution

We now turn to our comparative study of a Hamiltonian quantum evolution of a harmonic resonator (8.22) with various initial boundary conditions. Specifically, we find it illustrative to compare the evolution of Fock states to those of coherent states for low photon numbers, that is, $|n = 0, 1, 2\rangle$ versus $|\alpha| = 0, 1, 2$.

In Fig. 8.6, we show the time evolution of various coherent states calculated with eqn (8.32). Since the quantum harmonic oscillator is not in contact with any environment or drive, there is no energy decay and the oscillation continues with constant amplitude and phase and with a fixed particle number $n = |\alpha|^2 = x^2 + p^2$. The expectation values of x and p behave exactly as in the classical system. However, due to the uncertainty principle, there are inherent quantum fluctuations in position and momentum whose standard deviations are shown as green areas.

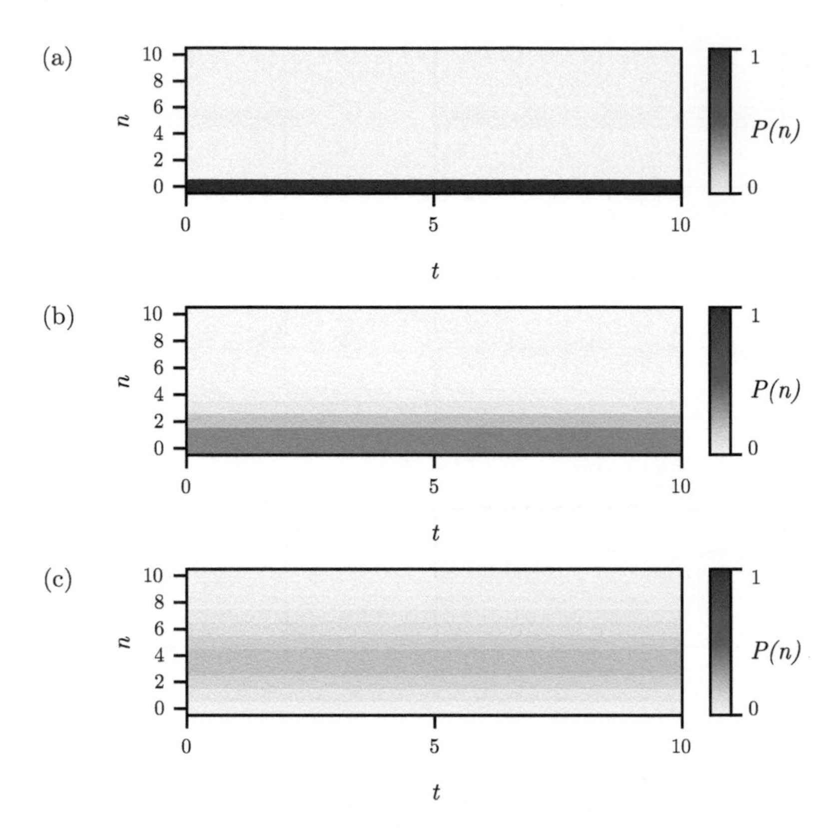

Fig. 8.8 Time evolution of the spectrum of Fock states $|n\rangle$ contributing to a quantum harmonic oscillator initialized in the coherent state (a) $|\alpha| = 0$, (b) $|\alpha| = 1$, and (c) $|\alpha| = 2$ without any driving or coupling to an environment. The relative probability $P(n)$ of each Fock state is represented by a color bar.

Even though the energy of the coherent state $|\alpha| = 1$ is the same as that of a Fock state $|n = 1\rangle$, the two states are strikingly different. This becomes apparent in Fig. 8.7, where the time evolution of the Fock states $|n = 0, 1, 2\rangle$ is shown. Here, the energy of the state is in no way related to the expectation values of x and p, which remain zero at all times. The difference in energy between for example $|n = 0\rangle$ and $|n = 2\rangle$ can only be seen in the increased standard deviation (green shaded area). In the following, we will employ complementary representations of the two states to better understand this difference.

In Figs 8.8 and 8.9, we show how the decomposition of three different states into their Fock spectrum evolves in time.[2] We clearly see how the coherent states are composed of several Fock states with their maximum amplitude $P(n)$ located around the index $n = |\alpha|^2$, while the pure Fock state $|n\rangle$ is, naturally, represented by a single component with unity probability, $P(n) = 1$. Again, no time evolution is

[2] Note that for a harmonic oscillator, the number operator commutes with the Hamiltonian, $[H, n] = 0$, such that this is equivalent to observing the time evolution of the eigenstates of the system [122].

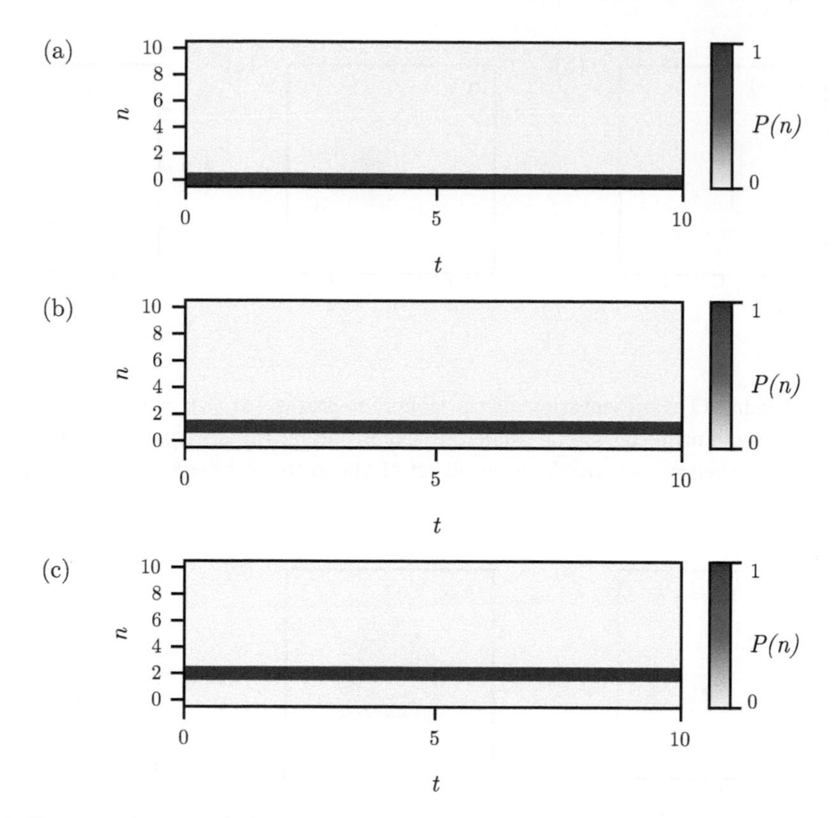

Fig. 8.9 Time evolution of the spectrum of Fock states $|n\rangle$ contributing to a quantum harmonic oscillator initialized in the Fock state (a) $|n=0\rangle$, (b) $|n=1\rangle$, and (c) $|n=2\rangle$ without any driving or coupling to an environment. The relative probability $P(n)$ of each Fock state is represented by a color bar.

apparent because the quantum harmonic oscillator experiences no external driving or dissipation.

In many cases, representations in phase space allow for a more intuitive understanding of a system. In Fig. 8.10, we see the Husimi-Q representation of the coherent states $|\alpha| = 0, 1, 2$, cf. Section 8.2.3. The state $|\alpha| = 0$ bears a striking resemblance to the thermal state in Fig. 5.4(a). However, the coherent state has a minimum uncertainty in both x and p that is a consequence of the Heisenberg uncertainty principle, not of classical force noise. The coherent states are shown here in a nonrotating frame, which means that they move in a circle around the origin as a function of time.

In Fig. 8.11, we visualize the lowest three Fock states in the Husimi-Q representation. While the Fock state $|0\rangle$ is identical to the coherent state $|\alpha| = 0$, the difference between Fig. 8.11(b) and (c) and Fig. 8.10(b) and (c) is obvious. Being eigenstates of the Hamiltonian H, Fock states have a well-defined energy but are maximally undefined in phase, such that their Husimi-Q images form a ring centered around the origin. This is the reason why the expectation values of x and p are zero at all times for pure Fock states, cf. Fig. 8.7.

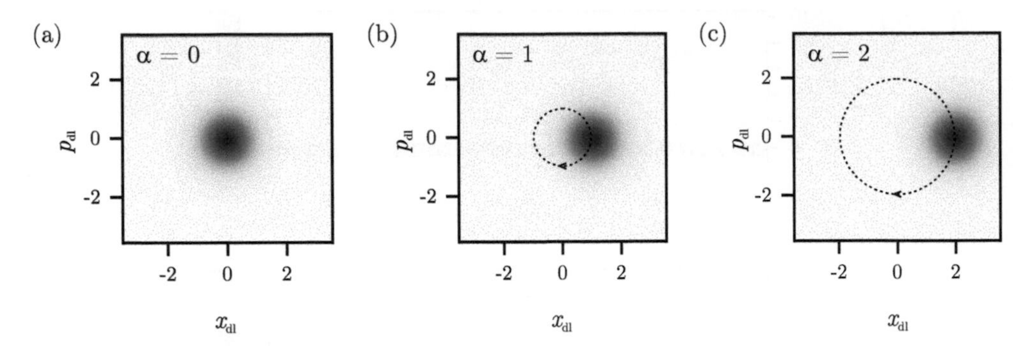

Fig. 8.10 Husimi-Q representation of the coherent states (a) $|\alpha| = 0$, (b) $|\alpha| = 1$, and (c) $|\alpha| = 2$ of a quantum harmonic oscillator. Color coding ranges from white for 0 to dark blue for $1/\pi$. Dashed arrows mark the rotation of the state in phase space as a function of time.

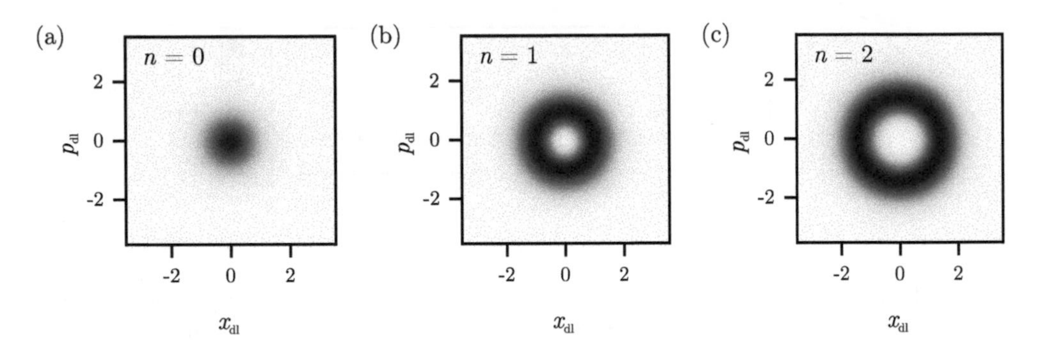

Fig. 8.11 Husimi-Q representation of the Fock states (a) $|n = 0\rangle$, (b) $|n = 1\rangle$, and (c) $|n = 2\rangle$ of a quantum harmonic oscillator in a nonrotating frame. Color coding ranges from white for 0 to dark blue for $1/\pi$.

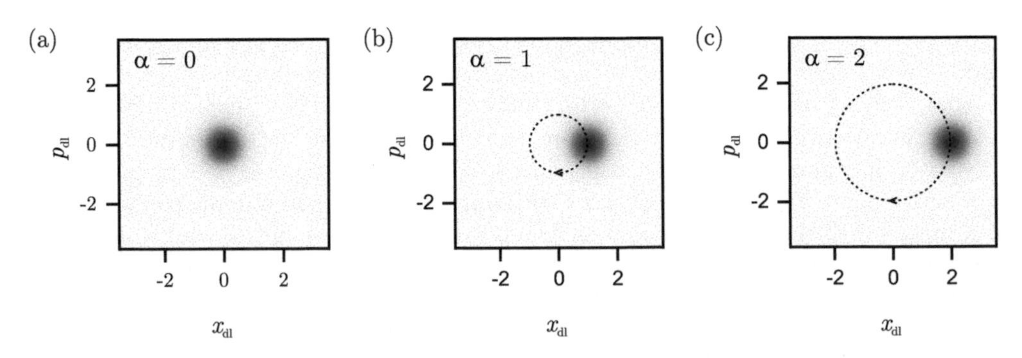

Fig. 8.12 Wigner representation of the coherent states (a) $|\alpha| = 0$, (b) $|\alpha| = 1$, and (c) $|\alpha| = 2$ of a quantum harmonic oscillator. Color coding ranges from dark red for $-1/\pi\hbar$ to white for 0 and to dark blue for $1/\pi\hbar$. Dashed arrows mark the rotation of the state in phase space as a function of time.

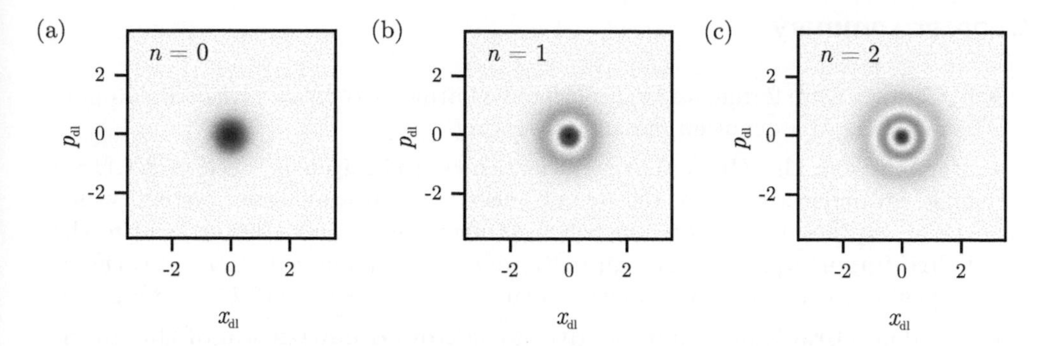

Fig. 8.13 Wigner representation of the Fock states (a) $|n = 0\rangle$, (b) $|n = 1\rangle$, and (c) $|n = 2\rangle$ of a quantum harmonic oscillator. Color coding ranges from dark red for $-1/\pi\hbar$ to white for 0 and to dark blue for $1/\pi\hbar$.

The Wigner quasiprobability distributions of the lowest three coherent states and Fock states are shown in Figs. 8.12 and 8.13. The main difference from the Husimi-Q representation is the appearance of phase space regions with negative values for non-classical states. As we can see, Fock states are distinctly nonclassical, as the notion of an oscillation that has no defined phase at any given time is unphysical in a classical setting. In contrast, coherent states can be generated through a combination of coherent drives and damping and have no negative regions in the Wigner quasiprobability distributions. Indeed, as we mentioned above, the coherent state is essentially the quantum description of a zero-temperature classical state.

In Chapter 9, we will rely on these introductory examples of quantum representations to study more complex states and time evolutions.

Chapter summary

- In Chapter 8, we formulate the problem of a harmonic oscillator without damping and driving in a quantum mechanical setting.
- Starting from the **Heisenberg uncertainty principle** in Section 8.1.1, we build an understanding of the role of uncertainty in a quantum system, which can be expressed as an intrinsic wave character. This discussion leads us to the **Schrödinger equation**, cf. eqn (8.6), and to the quantized eigenstates (**Fock states**) of the quantum harmonic oscillator, cf. eqns (8.9) to (8.11) and Fig. 8.4.
- We adopt **Dirac's notation** to arrive at a **second quantization of the quantum Hamiltonian**, cf. eqns (8.21) and (8.22).
- We introduce the **coherent state** as a useful basis for open quantum systems, cf. eqn (8.25). A coherent state resembles a classical state with quantum fluctuations instead of thermal noise.
- In order to deal with systems that cannot be described as a single wave function Ψ, we introduce the **density matrix** ρ, cf. eqn (8.28). The time evolution of density matrices cannot be solved with the Schrödinger equation. Instead, we rely on the **von Neumann master equation**, cf. eqn (8.32).
- We discuss a few useful representations and visualizations of quantum states in Section 8.3, such as the time evolution of $x(t)$ and $p(t)$ with corresponding uncertainties σ_x and σ_p, the **Fock spectrum**, and phase space representations like the **Husimi-Q distribution** and the **Wigner quasiprobability distribution**. Using these representations, we can visualize the differences between Fock states and coherent states.

Exercises

Check questions:

(a) Compare the properties of coherent states and Fock states to those of a classical harmonic oscillator in the presence of force noise. What features are similar? What features are clearly non-classical? In particular, what is the meaning of the negative regions in the Wigner representation?

(b) What is the difference between a classical and a quantum harmonic oscillator for $T \to 0$?

(c) How do the expectation values of a quantum oscillator's x and p compare to their classical counterparts?

(d) What distinctly non-classical features did we encounter in this chapter, and why can they not be observed in our everyday life?

(e) Discuss three- and four-wave mixing in a quantum mechanical setting, cf. Fig. 8.5.

Tasks:

8.1 Open the code **Python Example 8** and familiarize yourself with its features. Note that the simulation is based on the Lindblad master equation, which we will present in Chapter 9. For a closed system, that is, negligible damping ($\kappa = 0$) and zero temperature ($n_{\text{th}} = 0$), its results are equivalent to those of the von Neumann equation, cf. eqn (8.32).

8.2 Use $N = 22$ as the size of the truncated Hilbert space. Set $\kappa = n_{\text{th}} = 0$ to simulate a closed quantum system (the meaning of *closed* versus *open* will be discussed in detail in Chapter 9). Initialize the system in the coherent state $|\alpha| = 2$ and observe its time dependence in the provided representations. Which features of the results can you reproduce with classical theory from earlier chapters?

8.3 Repeat the experiment above with various coherent states. What do you observe when the state becomes too large ($\alpha > 3$)? What is this effect due to, and how can it be corrected for?

8.4 Initialize the system in various Fock states. Which features of the results can you reproduce with classical theory from earlier chapters?

8.5 Activate the time dependence of the Hamiltonian H_0 by selecting appropriate values for **h0** and **t0**. What happens to the system for various initial conditions? What is the meaning of the change in the Hamiltonian? Can you exploit this handle to create interesting states?

8.6 Repeat the previous exercise with the Hamiltonian H_1 and its parameters **h1** and **t1**. What is the difference from the case with H_0? Can this control knob be useful for quantum engineering?

9 From Closed to Open Quantum Systems

In Chapter 8, we reviewed some basics of quantum mechanics in the example of the quantum harmonic oscillator. We limited ourselves to a Hamiltonian description that is devoid of time dependencies such as damping or external driving. In this chapter, we open our system by allowing it to interact with elements beyond the limits of the closed Hamiltonian. These elements are responsible for explicit time dependencies by injecting energy into the oscillator, and by allowing energy to drain back into the environment. There are several textbooks dealing with open quantum theory [134, 136–138]. Our derivation here is presented succinctly and tailored to our specific needs.

9.1 Coupling to a Thermal Environment

We start by introducing the environment degrees of freedom in order to obtain a quantum mechanical description for damping. We introduced in Chapter 4 the environment as a white force noise PSD acting on the resonator, cf. eqn (4.5). This picture was based on the assertion that energy dissipation into an environment must be accompanied by force fluctuations in order to restore thermal equilibrium. This balance was finally formulated in terms of the FDT in Section 4.2. We assumed the environment to be infinitely large and unaffected by the system's back action, that is, the environment can accept energy from the system without changing its own temperature. We now wish to translate this description to quantum mechanics. Here, the white spectrum of the environment can be understood as an ensemble of harmonic oscillators that are coupled to the system. Each resonator in the environment exchanges energy with the system at its own eigenfrequency.

9.1.1 Liouville and Lindblad Master Equations

Until now, our description of quantum mechanical systems has relied on the Hamiltonian. Hence, to include the environment degrees of freedom, we start by writing the Hamiltonian description of the total setup

$$H_{\text{tot}} = H_{\text{sys}} + H_{\text{env}} + C_{\text{coupl}} H_{\text{coupl}} \,. \tag{9.1}$$

Here, H_{sys} describes our closed system, cf. eqn (8.22). H_{env} describes the environment degrees of freedom, which in the case of a harmonic oscillator ensemble reads

$$H_{\text{env}} = \sum_i \hbar\omega_i a_i^\dagger a_i \,. \tag{9.2}$$

To connect with the white spectrum of the classical environment, we assume ω_i to densely cover all energies, and that the environment operators are decoupled $\left[a_i, a_j^\dagger\right] = \delta_{ij}$. The last term H_{coupl} describes the coupling between the system and the environment with amplitude C_{coupl}, and contains products of operators acting on the system and the environment. Without losing any generality, the interaction term can be decomposed as

$$H_{\text{coupl}} = \sum_j S_j \otimes E_j, \tag{9.3}$$

with S_j and E_j describing the system and the environment, respectively. In the following, we consider linear coupling between the system operators $S_j \in \{a, a^\dagger\}$ and environment operators composed of the raising and lowering operators of the resonators in eqn (9.2), $E_j \in \left\{a_i, a_i^\dagger\right\}$.

The time evolution of the total system follows the von Neumann equation from eqn (8.32),

$$\dot{\rho}_{\text{tot}}(t) = -\frac{i}{\hbar}\left[H_{\text{tot}}, \rho_{\text{tot}}(t)\right], \tag{9.4}$$

where the density matrix ρ_{tot} describes the state of the total system. We are, however, mostly interested in the small subsystem that corresponds to our oscillator, which we refer to as *reduced system* or simply as *system* with the density matrix $\rho \equiv \rho_{\text{sys}} \equiv \text{tr}_{\text{env}}\{\rho_{\text{tot}}\}$. Note that the partial trace averages over all possible configurations of the environment and retains the system degrees of freedom. The evolution of the infinitely sized environment is of little concern for our purposes. The goal of open quantum theory is to infer the equations of motion of the reduced system from those of the total system. This procedure generally engenders **Liouville's equation of motion**

$$\dot{\rho}(t) = \mathcal{L}\rho, \tag{9.5}$$

where the Liouvillian superoperator $\mathcal{L}$ describes both the coherent time evolution of the system due to its own Hamiltonian, and the impact of the environment on the system.

For practical purposes, we expect that the reduced equations of motion should be easier to solve than the full dynamics of the total system. To fulfill this requirement, several approximations are made in the derivation of the reduced dynamics. Specifically, we invoke the following three assumptions that are usually applied: (i) the coupling between environment and system is weak, such that the system can maintain long coherence (high Q); (ii) excitations decay quickly in the environment, that is, the environment is infinitely large and has no memory of the energy it absorbs; and, relevant to the driven cases discussed here, (iii) only environment frequencies close to ω_0 allow resonant energy exchange and are important for the description of the system–environment coupling.

Under these assumptions, the coupling between system and environment can be introduced in the form of Lindblad operators that act on the density matrix ρ [137, 139]. This yields the **master equation in Lindblad form,**

$$\dot{\rho}(t) = -\frac{i}{\hbar}[H, \rho] + \kappa_i \sum_i \mathcal{D}[L_i]\rho\,, \tag{9.6}$$

where the Lindblad superoperator $\mathcal{D}$ is defined as

$$\mathcal{D}[L_i]\rho = L_i \rho(t) L_i^\dagger - \frac{1}{2}\left\{L_i^\dagger L_i, \rho(t)\right\}. \tag{9.7}$$

Here, the κ_i are system–environment coupling rates and the L_i are referred to as *jump operators* that are calculated as linear combinations of the S_j, cf. section B in Ref. [139].

In the following, we focus on the case where the jump operators are the raising and lowering operators of the system, $L_i \in \{a, a^\dagger\}$. Note that in experiments, dephasing of the form $L_{\text{deph}} \equiv a^\dagger a$ also plays a role [63] but is not addressed here. We include linear damping with the coupling coefficient κ, which is the rate at which the oscillator exchanges energy with the environment through the annihilation and creation operators a and $a^\dagger$. This form of including dissipation in a quantum system leads to

$$\dot{\rho}(t) = -\frac{i}{\hbar}[H, \rho] + \kappa(1 + n_{\text{th}})\mathcal{D}[a]\rho + \kappa n_{\text{th}}\mathcal{D}[a^\dagger]\rho\,. \tag{9.8}$$

Each of the oscillators in the environment must be in thermal equilibrium with the environment temperature T, which entails that the environment has a mean thermal excitation n_{th} at the oscillator's eigenfrequency ω that follows the Bose–Einstein distribution

$$n_{\text{th}}(\omega) = \frac{1}{e^{\frac{\hbar\omega}{k_B T}} - 1}\,. \tag{9.9}$$

In eqn (9.8), the terms including $\mathcal{D}[a^\dagger]$ and $\mathcal{D}[a]$ are responsible for transporting energy quanta from the environment to the oscillator and vice versa, respectively. Note that there are three different terms: $\kappa n_{\text{th}}\mathcal{D}[a^\dagger]$ and $\kappa n_{\text{th}}\mathcal{D}[a]$ describe thermally activated processes and correspond to the thermal energy fluctuations that, in the classical picture, are caused by the force noise ξ. These terms correspond to stimulated absorption and emission, respectively [134]. When the temperature of the environment is vanishingly small and $n_{\text{th}} \approx 0$, the only remaining term is $\kappa\mathcal{D}[a]$, which provides a unidirectional channel for energy flow from the oscillator to the environment. This is the quantum equivalent of the damping term $\Gamma\dot{x}$ that leads to amplitude saturation and ringdown in a classical EOM, cf. Chapter 1. This process is known as spontaneous emission [134].

In the classical treatment, we found that nonlinear damping is unimportant for most aspects of the parametric oscillator. Mainly, it was included to provide a termination of the high-amplitude branch of the parametric oscillator, creating the bifurcation that divides regions III and IV in Fig. 3.2. We do not address nonlinear damping in our description of the quantum parametric oscillator.

9.1.2 Effect of the Thermal Environment

Equation (9.8) supplies us with a general framework to follow the evolution of a system while it is coupled to a thermal environment. We begin by studying energy decay

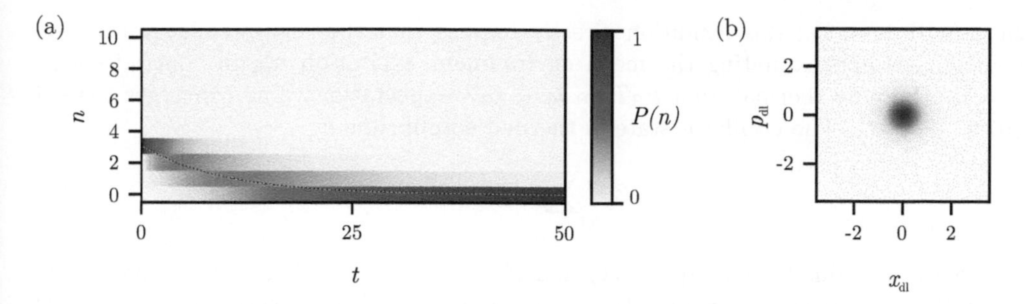

Fig. 9.1 Energy decay from a Fock state $|3\rangle$ into a cold environment, with $m = 1$, $\omega_0 = 1$, $\kappa = 0.1$, and $\hbar = 1$. The environment has a mean excitation number $n_{\text{th}} = 0$, corresponding to a temperature $T = 0$. (a) Time evolution of the Fock spectrum. A black line indicates $\langle n \rangle$, while a blue dashed line (exactly on top of the black one) marks the classical exponential decay $e^{-\Gamma t}$ with $\Gamma = \kappa$. (b) Wigner distribution of the oscillator state at $t = 50$.

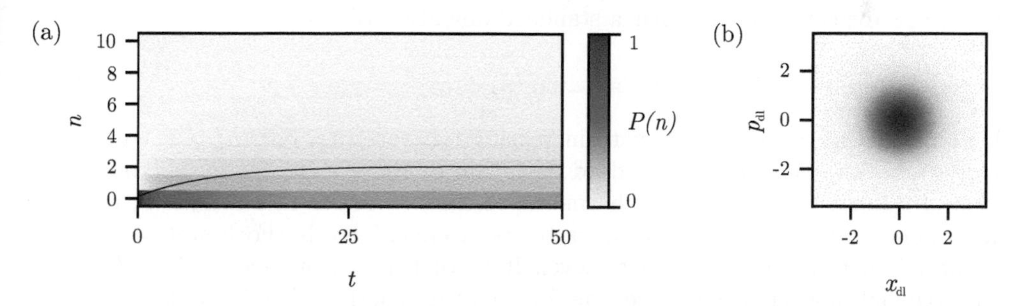

Fig. 9.2 Effect of a thermal environment on a quantum harmonic oscillator initialized in $|0\rangle$, with $m = 1$, $\omega_0 = 1$, $\kappa = 0.1$, and $\hbar = 1$. The environment has a mean excitation number $n_{\text{th}} = 2$. (a) Time evolution of the Fock spectrum. A black lines indicates $\langle n \rangle$. (b) Wigner distribution of the oscillator state at $t = 50$.

from the quantum harmonic oscillator into a cold environment. In Fig. 9.1, we plot the solution of eqn (9.8) for an oscillator initialized in the excited state $|3\rangle$ and linearly coupled to an environment at a temperature $T = 0$, that is, with a mean thermal excitation $n_{\text{th}} = 0$. In this situation, two of the three Lindblad superoperators in eqn (9.8) vanish and the only dynamics we observe is energy transport from the oscillator to the environment, causing an exponential oscillation decay. Once the oscillator has reached its ground state, the system does not evolve anymore. This is the quantum analog of the ringdown in Fig. 1.4, where we effectively also modeled damping as a unilateral energy flow with rate Γ. It is therefore not surprising that the decrease of $\langle n \rangle$ can be described by the classical exponential decay term, as shown by a blue dashed line.

In Fig. 9.2, we observe the evolution of an oscillator initialized in its ground state $|0\rangle$ and coupled to an environment at a finite temperature T. The linear coupling κ causes bilateral energy exchange between the environment and the oscillator, such that the pure Fock state $|0\rangle$ evolves into a thermal mixture of Fock states [122, 140]. A thermal equilibrium is reached when the oscillator receives as much energy from the environment as it sends back. Since the relevant oscillators in the environment are those with $\omega \approx \omega_0$, this condition directly implies that the subsystem's expectation value $\langle n \rangle$ is approximating the mean environment excitation n_{th} in eqn (9.9) with $\omega = \omega_0$. For the thermal limit $k_B T \gg \hbar\omega_0$, this expectation value converges toward $\langle n \rangle \approx \frac{k_B T}{\hbar\omega_0} - \frac{1}{2}$. The oscillator state in thermal equilibrium is

$$\rho = Z^{-1} \sum_n e^{\frac{-E_n}{k_B T}} |n\rangle \langle n| \,, \tag{9.10}$$

where E_n is defined as in eqn (8.11) and $Z = \text{tr}(e^{-H/k_B T})$ is a normalization factor [122]. The expectation value of the oscillator energy follows directly from eqns (8.11) and (9.9) as

$$\langle H \rangle = \hbar\omega_0 \left(\frac{1}{2} + \langle n \rangle \right) = \frac{\hbar\omega_0}{2} + \frac{\hbar\omega_0}{e^{\hbar\omega_0} k_B T - 1} \,, \tag{9.11}$$

converging toward $\langle H \rangle \approx \hbar\omega_0 \langle n \rangle$ in the thermal limit. The Bose–Einstein distribution around this expectation value has a standard deviation of [141]

$$\sigma_E = \hbar\omega_0 \sqrt{\langle n \rangle + \langle n \rangle^2} \,, \tag{9.12}$$

which approaches $\sigma_E \approx \hbar\omega_0 \langle n \rangle$ with increasing temperature. For $\langle n \rangle > 1$, eqn (9.12) yields $\sigma_E > \hbar\omega_0$, that is, the standard deviation of the system energy is larger than the level spacing of the quantum harmonic oscillator. Under this condition, thermal fluctuations induce spontaneous jumps between energy levels (Fock states) and the oscillator is not confined to a single level. It is for this reason that $k_B T \ll \hbar\omega_0$ is a useful (though not sufficient) condition for quantum control of oscillators.

9.2 The Driven Quantum Resonator

After having established how energy is exchanged between a quantum harmonic oscillator and a thermal environment, we address how external driving can be included in our system. To this end, we first take a look at rotating-frame methods that allow us to study a system's slowly varying quadratures, similar to the classical van der Pol transformation in eqn (2.22).

9.2.1 The Rotating Phase Space

When treating classical stochastic systems, we found it convenient to study our results in a rotating phase space. Similarly, with a quantum system, one commonly enters a frame rotating at ω (e.g. at the frequency of an external drive) by using the transformation

$$\tilde{H} \equiv U_{\text{rot}}^\dagger H U_{\text{rot}} - i\hbar U_{\text{rot}}^\dagger \dot{U}_{\text{rot}} \tag{9.13}$$

with the unitary operator[1] $U_{\text{rot}}(t) = e^{-i\omega t a^\dagger a}$. Generally, the rotating Hamiltonian $\tilde{H}$ will have terms proportional to the detuning $\Delta = \omega - \omega_0$ as well as so-called

[1] This unitary transformation corresponds to a counter-clockwise rotation in phase space, $x \propto e^{i\omega t}$, while for the classical treatment we employed a clockwise rotation, $x \propto e^{-i\omega t}$. The v axis is therefore inverted between the classical and quantum rotating phase spaces.

micromotion terms at $\omega + \omega_0$. The latter oscillate rapidly in the new frame and are therefore discarded, leaving only the terms at Δ. This procedure is called the **rotating-wave approximation** (RWA). It is often valid under the condition $|1 - \frac{\omega_0}{\omega}| \ll 1$ and when the micromotion terms are small enough.

For the Hamiltonian in eqn (8.22), we obtain the rotating Hamiltonian

$$\tilde{H} = -\hbar\Delta\tilde{a}^\dagger\tilde{a} \tag{9.14}$$

with the rotated operator $\tilde{a} \equiv U_{\mathrm{rot}}^\dagger a U_{\mathrm{rot}}$. This operator can be used to define approximate slow-flow operators u and v in phase space,

$$u \approx \sqrt{\frac{\hbar}{2m\omega_0}}(\tilde{a}^\dagger + \tilde{a}), \tag{9.15}$$

$$v \approx -i\sqrt{\frac{\hbar}{2m\omega_0}}(\tilde{a}^\dagger - \tilde{a}), \tag{9.16}$$

and correspondingly

$$\tilde{a} \approx (u + iv)\sqrt{\frac{m\omega_0}{2\hbar}}, \qquad \tilde{a}^\dagger \approx (u - iv)\sqrt{\frac{m\omega_0}{2\hbar}}. \tag{9.17}$$

For the expectation values $\langle u\rangle$, $\langle v\rangle$, $\mathrm{Re}\,\langle\tilde{a}\rangle$, and $\mathrm{Im}\,\langle\tilde{a}\rangle$ we find the relations

$$\langle u\rangle \approx \sqrt{\frac{2\hbar}{m\omega_0}}\,\mathrm{Re}\,\langle\tilde{a}\rangle \equiv \sqrt{\frac{2\hbar}{m\omega_0}}\alpha_R, \qquad \langle v\rangle \approx -\sqrt{\frac{2\hbar}{m\omega_0}}\,\mathrm{Im}\,\langle\tilde{a}\rangle \equiv \sqrt{\frac{2\hbar}{m\omega_0}}\alpha_I, \tag{9.18}$$

where we introduced the symbols α_R and α_I for the real and imaginary quadratures of the complex amplitude $\tilde{\alpha} = \langle\tilde{a}\rangle$, respectively. In the following, we will present Wigner quasiprobability distributions in terms of these dimensionless quadratures, which are approximately proportional to[2] $\langle u\rangle$ and $\langle v\rangle$. The Wigner quasiprobability distributions in this rotating phase space can be calculated according to the formula[142]

$$W(\alpha_R, \alpha_I) = \frac{1}{\pi}\int_{-\infty}^{\infty} \langle\alpha_R + \alpha_R'|\,\rho\,|\alpha_R - \alpha_R'\rangle\, e^{-i2\alpha_I\alpha_R'}d\alpha_R'\,. \tag{9.19}$$

The above approximation of small detuning from resonance delivers results with satisfying accuracy for many applications, for example in quantum optics, where the relevant detuning is typically very small compared to the bare resonance frequency. In other fields, this is no longer generally true, and alternative methods provide more

[2] The proportionality is exact for $\Delta = 0$.

accurate descriptions of a system in the rotating frame. One option is to define new operators that describe the system in terms of excitations not at the eigenfrequency ω_0, but at a general frequency ω, which can be the frequency of an external drive [143]. These operators take the form

$$b = \sqrt{\frac{m\omega}{2\hbar}}\left(x + \frac{i}{m\omega}p\right), \qquad (9.20)$$
$$b^\dagger = \sqrt{\frac{m\omega}{2\hbar}}\left(x - \frac{i}{m\omega}p\right),$$

with the inverse transformation

$$x = \sqrt{\frac{\hbar}{2m\omega}}\left(b^\dagger + b\right), \qquad (9.21)$$
$$p = i\sqrt{\frac{\hbar m\omega}{2}}\left(b^\dagger - b\right),$$

and analogous formulations for u and v, cf. eqn (9.15). In this new notation, the Hamiltonian of the quantum harmonic oscillator takes the form

$$H = \hbar\frac{\omega_0^2 + \omega^2}{2\omega}\left(b^\dagger b + \frac{1}{2}\right) + \hbar\frac{\omega_0^2 - \omega^2}{4\omega}\left(b^2 + (b^\dagger)^2\right). \qquad (9.22)$$

While this Hamiltonian is less straightforward than its counterpart in eqn (8.22), the additional terms ensure that the results obtained, in a frame rotated by $U_{\text{rot}}(t) = e^{-i\omega t b^\dagger b}$, are correct also in the case of, for example, a detuned drive [143]. Nevertheless, we will in the following stick to the well-known description in terms of $a^\dagger$ and a and the respective RWA form to facilitate a direct comparison with results in the larger part of the literature.

9.2.2 External Driving

We are now ready to include an external drive into our quantum description. At this point, the Hamiltonian we consider is

$$H = \frac{p^2}{2m} + \frac{1}{2}m\omega_0^2 x^2 - F_0\cos(\omega t + \theta)x, \qquad (9.23)$$

with $\omega \approx \omega_0$. We can apply eqn (8.21) to these terms and obtain in the notation of second quantization the equivalent Hamiltonian

$$H = \hbar\omega_0(a^\dagger a + 1/2) - F_0\sqrt{\frac{\hbar}{2m\omega_0}}\cos(\omega t + \theta)(a^\dagger + a). \qquad (9.24)$$

In the RWA for $|F| \ll \Delta$, eqn (9.24) is transformed into

$$\tilde{H} = -\hbar\Delta\tilde{a}^\dagger\tilde{a} - F(\tilde{a}^\dagger e^{-i\theta} + \tilde{a}e^{i\theta}), \qquad (9.25)$$

with

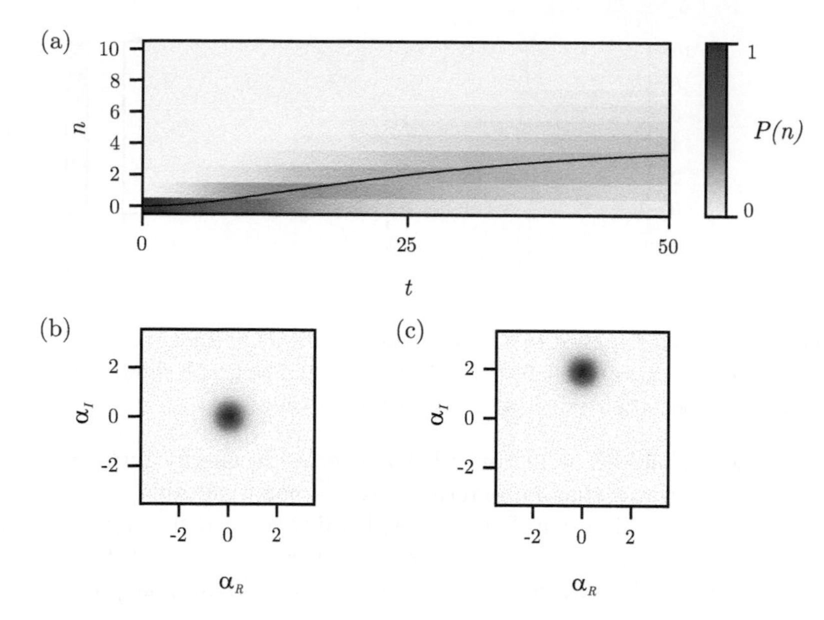

Fig. 9.3 External force applied to a quantum harmonic oscillator initialized in $|0\rangle$, with $m = 1$, $\omega = \omega_0 = 1$, $\kappa = 0.1$, $F = 0.1$, $\theta = 0$, and $\hbar = 1$. The environment has a mean excitation number $n_{\mathrm{th}} = 0$. (a) Time evolution of the Fock spectrum. A black line indicates $\langle n\rangle$. (b) Wigner distribution of the oscillator state in the rotating phase space at $t = 0$, cf. eqn (9.19). (c) Wigner distribution of the oscillator state at $t = 50$.

$$F = \frac{F_0}{2}\sqrt{\frac{\hbar}{2m\omega_0}}. \tag{9.26}$$

We find that in this description, the external force corresponds to the simultaneous application of both operators $\tilde{a}^\dagger$ and $\tilde{a}$ with different phases.[3] Does this mean that the driven quantum harmonic oscillator state simply ascends along Fock states?

In Fig. 9.3, we show the result of eqn (9.8) for the Hamiltonian in eqn (9.25). Starting from the ground state, we observe an increase of the mean energy $\langle n\rangle$ in response to the drive which increases the corresponding uncertainty σ_E, see Fig. 9.3(a). The results beyond $t = 0$ are thus not simple Fock states, even for a zero-temperature environment. At the same time, the uncertainties of x and p remain the same between the initial and final states, see Fig. 9.3(b) and (c). The drive merely displaces the ground state, which is then described as a coherent state $\tilde{\alpha}$. This observation is in accordance with the **correspondence principle**, which demands that the well-known results of the classical driven and damped harmonic oscillator must be described by the quantum model for large $\langle n\rangle$.

We repeat the numerical simulation with a fully isolated driven oscillator ($\kappa = 0$) in Fig. 9.4. The absence of dissipation allows for unlimited growth of the amplitude,

[3] Please note that we reuse some symbols with a new meaning in the quantum treatment relative to the classical terminology in order to use a typical notation.

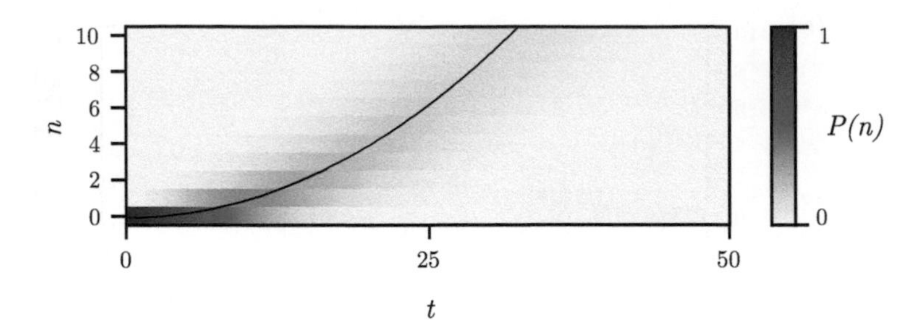

Fig. 9.4 External force applied to a fully isolated quantum harmonic oscillator initialized in $|0\rangle$, with $m = 1$, $\omega = \omega_0 = 1$, $\kappa = 0$, $F = 0.1$, and $\hbar = 1$. The graph shows the Fock spectrum and occupation expectation value $\langle n \rangle$ as a function of time.

but it is obvious that the oscillator still has a growing energy uncertainty σ_E along with $\langle n \rangle$. This indicates that an external force is not a suitable tool to prepare the quantum harmonic oscillator in Fock states, leading instead to amplitude eigenstates α that are the closest quantum equivalent to classical oscillator states.

Given that our quantum description relies on individual, sharp energy levels, we could expect that the drive only has an effect for $\Delta = 0$. As soon as there is a detuning between the level separation and the frequency of the drive, the process visualized in Figs 9.3 and 9.4 should naively be forbidden. At this point, however, we recall the Mandelshtam–Tamm inequality in eqn (8.2), which asserts that a state with a finite lifetime σ_τ has an uncertainty in its energy of $\sigma_E \geq \frac{\hbar}{2\sigma_\tau}$. This uncertainty leads to an effective energy broadening of the allowed transitions between levels, such that a detuned force can still drive the oscillator. In our damped quantum harmonic oscillator, the lifetime is given by the decay of energy to the environment with a rate $\kappa = \Gamma$ [124]. The quantum treatment thus provides a new interpretation of the connection between the damping rate (Γ) and the FWHM ($\Delta\omega$) of a classical resonator that we introduced in Section 1.2.

9.2.3 Equation of Motion for $\tilde{\alpha}$

Sometimes we are not interested in the behavior of the entire density matrix ρ, but only in the expectation value $\tilde{\alpha} = \langle \tilde{a} \rangle$ of a coherent state. We may then reformulate eqn (9.8) in terms of $\tilde{\alpha}$ using $\tilde{\alpha} = \mathrm{tr}(\tilde{a}\rho)$ from eqn (8.34). With this definition and using that $\frac{d}{dt}\mathrm{tr}(\tilde{a}\rho) = \mathrm{tr}(\tilde{a}\dot{\rho})$, the master equation becomes for $T = 0$

$$\dot{\tilde{\alpha}} = i\Delta\tilde{\alpha} + i\frac{F}{\hbar}e^{-i\theta} - \frac{\kappa\tilde{\alpha}}{2}\,. \tag{9.27}$$

This procedure of combining a mean-field ansatz (studying $\tilde{\alpha}$ instead of $\tilde{a}$) with the Lindblad approximations is useful to understand the similarities between quantum and classical systems, and will be picked up again in Chapter 10. At the same time, it is important to be aware of the limitations of eqn (9.27): the equation is based on the RWA, and therefore yields a linear approximation to the quadratic response curve in eqn (1.13). For $\Delta \ll \kappa$, this approximation is acceptable, but it becomes inaccurate for $\Delta > \kappa$ because counter-rotating force terms are neglected [143].

Chapter summary

- In Chapter 9, we generalize our quantum model to describe damped and driven quantum harmonic oscillators.
- We describe a thermal environment as an infinite number of harmonic oscillators covering all frequencies. Following a number of assumptions and approximations, we then arrive at the **Lindblad master equation**, cf. eqn (9.8). This equation replaces the von Neumann equation for a system in contact with an environment.
- Equation (9.8) allows us to describe energy decay from an oscillator to a cold environment, as well as thermalization of an oscillator in contact with a hot environment. We can therefore translate the physics discussed in Chapter 4 into our quantum model. The three terms containing superoperators in the Lindblad master equation correspond to stimulated absorption and emission, as well as spontaneous emission. They describe in the notation of second quantization the same processes that we previously associated, respectively, with stochastic fluctuations induced by the force noise ξ and with unilateral energy decay via Γ.
- In Section 9.2, we formulate a rotating-frame description of the quantum system, in analogy to the van der Pol transformation we used for classical systems. For most of this text, we rely on the well-known **rotating-wave approximation** for this purpose.
- We introduce an external drive into our quantum Hamiltonian, cf. eqns (9.24) and (9.25).
- Finally, we briefly show how an equation of motion for the expectation value $\tilde{\alpha}$ of a coherent state can be defined. Equation (9.27) is useful when a full description of the density matrix is not necessary.

Exercises

Check questions:

(a) In what case do you expect to see a clear difference between the steady state resulting from the Lindblad master equation and the corresponding fluctuating classical result obtained with for example the Fokker–Planck treatment? When will they look very similar?

(b) Both thermal and quantum fluctuations act on the oscillator. In an experiment, how can you distinguish between the two phenomena?

(c) What is the meaning of the mean excitation number n_{th}? How does it reach a stable value in an environment, and how is it translated to the system of interest?

(d) Why can an external force pulse not be used to create pure Fock states in the quantum harmonic oscillator?

Tasks:

9.1 Open the code **Python Example 9** and familiarize yourself with its features. Use $N = 22$ as the size of the truncated Hilbert space and select $\Delta = 0$ by setting "**Del_U=0**",[4] as well as $U = G = F = \kappa = 0$ and $n_{\text{th}} = 0$. Initialize the oscillator in various Fock states and observe the result in the provided representations.

9.2 Initialize the oscillator in $|n = 3\rangle$ and let the system evolve in the presence of a loss term $\kappa = 0.01$. Repeat the same numerical experiment starting from a coherent state with $|\tilde{\alpha}| = 3$. When comparing the decay processes, what differences do you notice, in particular with regard to the expectation values and variances of α_R and α_I?

9.3 Initialize the system in $|n = 0\rangle$ and drive it with $F = 0.1$ in the presence of $\kappa = 0.1$. The resulting dynamics should look familiar from Chapter 1. What formula describes the expectation values of α_R and α_I as a function of time? Can you interpret it in terms of a quantum mechanical picture?

9.4 Repeat the driving experiment a couple of times with various values of the detuning Δ included as **Del_U**. Compare the steady-state amplitudes α_R and α_I that your system reaches to the analytical result expected from Section 1.2 (with arbitrary units) and to the result from eqn (9.27). Can you explain what you observe?

9.5 Use the switching times $(\mathbf{t1}, \mathbf{t2}, \ldots)$ to switch the force on and off at selected points in time. Can you generate the Fock state $|n = 3\rangle$ in this way?

9.6 Play with the angle θ. What is the meaning of the angle in the phase plot, that is, what is the difference between a steady state centered on the u-axis and one on the v-axis?

9.7 Add a finite temperature to the system by using $n_{\text{th}} > 0$ and initialize the system again in $|n = 0\rangle$. First observe the role of this fluctuation force for $F = 0$ and $\kappa = 0.01$. Can you reduce the average thermal occupation of the oscillator by increasing κ? Then monitor the uncertainties in α_R and α_I as you apply an external drive F. Explain your observations based on what we learned in Chapter 4.

[4] The meaning of the index will become clear in Chapter 10.

10 The Quantum Parametric Oscillator

The exploration of dissipation and external driving in Chapter 9 prepared us for the next step, which is the addition of a parametric pump and a nonlinear potential term.[1] We will study the evolution of the system in the presence of these terms in a variety of situations to gain an understanding of their role in the quantum limit.

10.1 General Hamiltonian

We start from the classical EOM with $\eta = 0$, cf. eqn (3.1),

$$\ddot{x} + \omega_0^2(1 - \lambda\cos(2\omega t))x + \beta x^3 + \Gamma\dot{x} = \frac{F_0}{m}\cos(\omega t + \theta)\,, \tag{10.1}$$

where we set the parametric phase $\psi = 0$ for simplicity and without loss of generality. Ignoring for a moment the damping term, this EOM corresponds to the classical Hamiltonian

$$H = \frac{p^2}{2m} + \frac{m\omega_0^2}{2}(1 - \lambda\cos(2\omega t))x^2 + \frac{m\beta}{4}x^4 - F_0\cos(\omega t + \theta)x\,. \tag{10.2}$$

We can apply eqn (8.21) to these terms and obtain in the notation of second quantization the quantum Hamiltonian for the Kerr parametric oscillator,

$$\begin{aligned} H =& \hbar\omega_0(a^\dagger a + 1/2) - \frac{\hbar\omega_0\lambda}{4}\cos(2\omega t)\left(a^{\dagger 2} + a^2 + 2a^\dagger a + 1\right) \\ &+ \frac{\hbar^2\beta}{16m\omega_0^2}(a^\dagger + a)^4 - F_0\sqrt{\frac{\hbar}{2m\omega_0}}\cos(\omega t + \theta)(a^\dagger + a)\,. \end{aligned} \tag{10.3}$$

In the RWA, eqn (10.3) becomes

$$\tilde{H} = -\hbar\Delta_U\tilde{a}^\dagger\tilde{a} + \frac{U}{2}\tilde{a}^{\dagger 2}\tilde{a}^2 - F\left(\tilde{a}^\dagger e^{-i\theta} + \tilde{a}e^{i\theta}\right) - \frac{G}{2}\left(\tilde{a}^{\dagger 2} + \tilde{a}^2\right)\,, \tag{10.4}$$

with the expressions[2]

[1] In line with the typical terminology used in quantum optics, we refer to the nonlinear potential here as Kerr term instead of Duffing nonlinearity.

[2] Please remember that some symbols are reused and receive a new meaning in the quantum formalism.

$$\Delta_U = \Delta - U/\hbar\,, \tag{10.5}$$

$$U = \frac{3}{4}\frac{\hbar^2\beta}{m\omega_0^2}\,, \tag{10.6}$$

$$G = \frac{\hbar\omega_0}{4}\lambda\,, \tag{10.7}$$

denoting a detuning shifted by the nonlinearity, the Kerr nonlinear term, and the parametric pumping term, respectively. The Hamiltonian in eqn (10.4) can be inserted into the Lindblad master equation in eqn (9.8). This is our general strategy to obtain the time evolution of the quantum parametric oscillator.

10.1.1 Semi-Classical Equation of Motion

Following the procedure in Section 9.2.3, we can formulate an equation of motion for $\tilde{\alpha} = \langle\tilde{a}\rangle$ for the general nonlinear parametric oscillator. With all terms from eqn (10.4) included, eqn (9.27) becomes

$$\dot{\tilde{\alpha}} = i\Delta_U\tilde{\alpha} - i\frac{U}{\hbar}\tilde{\alpha}^*\tilde{\alpha}^2 + i\frac{F}{\hbar}e^{-i\theta} + i\frac{G}{\hbar}\tilde{\alpha}^* - \frac{\kappa\tilde{\alpha}}{2}\,, \tag{10.8}$$

with * denoting complex conjugation and where we used the semi-classical approximation $\langle(\tilde{a}^\dagger)^n\tilde{a}^m\rangle = (\langle\tilde{a}\rangle^*)^n\langle\tilde{a}\rangle^m$. The basis of coherent states is especially well suited to solve the semi-classical limit of large occupation $\langle n\rangle$. It is therefore interesting to compare the results of eqn (10.8) with those of the classical slow-flow equation. We insert eqn (9.18) into eqn (9.8), convert to the original physical parameters, and arrive at

$$\begin{aligned}-2\langle\dot{u}\rangle &= \left(2\omega_0 - 2\omega + \frac{\omega_0\lambda}{2}\right)\langle v\rangle + \frac{3}{4\omega_0}\langle X^2\rangle\beta\langle v\rangle + \kappa\langle u\rangle\\ &\quad - \frac{F_0}{m\omega_0}\sin(\theta) + \frac{3}{8}\frac{\hbar}{m\omega_0^2}\beta\langle v\rangle\,,\end{aligned} \tag{10.9}$$

$$\begin{aligned}-2\langle\dot{v}\rangle &= \left(-2\omega_0 + 2\omega + \frac{\omega_0\lambda}{2}\right)\langle u\rangle - \frac{3}{4\omega_0}\langle X^2\rangle\beta\langle u\rangle + \kappa\langle v\rangle\\ &\quad + \frac{F_0}{m\omega_0}\cos(\theta) - \frac{3}{8}\frac{\hbar}{m\omega_0^2}\beta\langle v\rangle\,,\end{aligned} \tag{10.10}$$

where in the rotating frame $\langle X^2\rangle = \langle u^2 + v^2\rangle = \frac{2\hbar}{m\omega_0}(|\tilde{\alpha}|^2 + 1/2)$. Inserting $\hbar \to 0$, $\kappa = \Gamma$, $\frac{\omega_0}{\omega} = 1 + O(\varepsilon)$, assuming that all system parameters are of order ε, and for $\langle u\rangle \to u$, $\langle v\rangle \to v$, and $\langle X^2\rangle \to X^2$ (up to an error $O(\varepsilon^2)$), we obtain the corresponding classical slow-flow equations

$$\dot{u} = -\frac{1}{2\omega}\left[\left(\omega_0^2 - \omega^2 + \frac{\omega_0^2\lambda}{2}\right)v + \frac{3}{4}X^2\beta v + \omega\Gamma u - \frac{F_0}{m}\sin(\theta)\right]\,, \tag{10.11}$$

$$\dot{v} = -\frac{1}{2\omega}\left[\left(\omega^2 - \omega_0^2 + \frac{\omega_0^2\lambda}{2}\right)u - \frac{3}{4}X^2\beta u + \omega\Gamma v + \frac{F_0}{m}\cos(\theta)\right]\,. \tag{10.12}$$

Equations (10.11) and (10.12) are identical to those we found for the deterministic description of the parametric resonator, cf. eqns (3.2) and (3.2). This means that in the semi-classical limit $\langle n \rangle \gg 1$, the coherent state of a nonlinear parametric quantum oscillator behaves like its classical counterpart — as we should expect from the correspondence principle. By contrast, for small particle numbers $\langle n \rangle \approx 1$, there can be important differences between classical and quantum systems. To explore these differences, we will in the following study various aspects of the nonlinear quantum parametric resonator.

10.2 Quantum Parametric Phenomena

In this section, we aim to understand the quantum parametric oscillator and how it relates to its classical analog which we studied in Chapters 3 and 5. As we have seen in Section 10.1.1, it is often not necessary to solve the quantum system's response at large amplitudes, as these approximate the classical dynamics. We can therefore concentrate on the solutions at low particle numbers n.

10.2.1 The Nonlinear Quantum Oscillator

We begin with the role of the nonlinearity in a quantum setting. Thus, for $G = 0$, let us find out how the driven and damped quantum oscillator is modified in the presence of a nonlinearity U.

In eqn (10.4), the rotating potential energy of the nonlinear quantum oscillator is described by the two terms

$$-\hbar\Delta_U \tilde{a}^\dagger \tilde{a} + \frac{U}{2}\tilde{a}^{\dagger 2}\tilde{a}^2 . \tag{10.13}$$

The first of the two terms in eqn (10.13) corresponds to a Fock ladder with equidistant states, that is, each particle corresponds to an added energy of $\hbar(\omega_0 + U)$ on top of a ground-state energy of $\hbar\omega_0/2$, cf. eqn (10.3). In a frame defined by the transformation $U_{\text{rot}}(t) = e^{-i\omega t a^\dagger a}$, we perceive this term only as a relative energy $-\hbar\Delta_U$ per particle, see Fig. 10.1.

The second of the two terms from eqn (10.13) indicates the nonlinear potential contribution. Like in the classical case, we work with an effective quartic potential whose potential energy depends on the displacement as $(a^\dagger + a)^4 \propto x^4$. In the RWA, this nonlinear term is simplified to $\frac{U}{2}\tilde{a}^{\dagger 2}\tilde{a}^2$. In the rotating Fock basis, this correction leads to different energy gaps $E_n - E_{n-1} = U(n-1)$, which for the state with $n = 2$ results in a nonlinear energy correction equal to $E_2 - E_1 = U$.

In a nonlinear resonator, an external force at $\Delta_U = 0$ is resonant with the lowest transition between the levels with $n = 0$ and $n = 1$. Beyond $n = 1$, however, this frequency no longer matches the energy differences between successive levels due to the presence of the nonlinear potential contribution U. In the presence of a strong nonlinearity $U \gg \kappa$, this nonlinear contribution exceeds the lifetime broadening of the Fock levels and the Fock ladder is effectively reduced to a **two-level system**.

The effect of an external force (field) on a two-level system is a textbook example of quantum mechanics [122, 138]. We illustrate it here in the example of a system with negligible spontaneous decay, that is, $\kappa = 0$. We write the state of the two-level system as

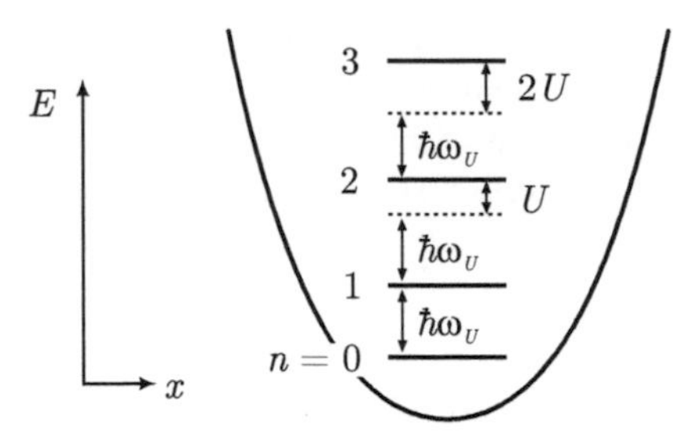

Fig. 10.1 Illustration of Fock states (levels) in a nonlinear potential. The nonlinearity introduces a shift of the equidistant contribution $\hbar\omega_U = \hbar(\omega_0 + U)$ as well as an additional energy correction that depends on the particle number n. For $n = 1$, the additional correction is zero, while for $n = 2$ it is equal to U. Due to this correction, the levels are no longer equidistant.

$$|\Psi\rangle = c_0 |0\rangle + c_1 |1\rangle , \tag{10.14}$$

where the time-dependent coefficients $|c_i|^2 = P_i$ indicate the probabilities of finding the system in the corresponding level. For a two-level system with an external force with phase $\theta = 0$, the Hamiltonian in eqn (10.4) simplifies to

$$\tilde{H} = -\hbar\Delta_U \tilde{a}^\dagger \tilde{a} - F\left(\tilde{a}^\dagger + \tilde{a}\right) , \tag{10.15}$$

where the ladder operators are only valid within the truncated Hilbert space spanned by $|0\rangle$ and $|1\rangle$. The Schrödinger equation for this system in matrix form is just

$$\frac{\partial}{\partial t}\begin{bmatrix} c_0 \\ c_1 \end{bmatrix} = -\frac{i}{\hbar}\tilde{H}\begin{bmatrix} c_0 \\ c_1 \end{bmatrix} = -i\begin{bmatrix} 0 & F/\hbar \\ F/\hbar & -\Delta_U \end{bmatrix}\begin{bmatrix} c_0 \\ c_1 \end{bmatrix} . \tag{10.16}$$

Differentiating eqn (10.16) with respect to time, we obtain the two decoupled equations [138]

$$\left(\frac{\partial^2}{\partial t^2} - i\Delta_U\frac{\partial}{\partial t} + \frac{F^2}{\hbar^2}\right) c_0 = 0 , \tag{10.17}$$

$$\left(\frac{\partial^2}{\partial t^2} - i\Delta_U\frac{\partial}{\partial t} + \frac{F^2}{\hbar^2}\right) c_1 = 0 . \tag{10.18}$$

For $\Delta_U = 0$, these two equations describe harmonic oscillations of both c_0 and c_1 as the system cycles between them, similar to what we obtained with parametric coupling in Section 6.2.1. Starting in, for example, $|n = 0\rangle$ corresponds to the initial conditions $P_0 = |c_0|^2 = 1$ and $P_1 = |c_1|^2 = 0$. A resonant force leads to a decrease of P_0 and an increase of P_1 over time. Once the system has reached $P_1 = 1$ and $P_0 = 0$, the drive brings the system back to $|n = 0\rangle$ since no higher states are available. The two directions of energy transfer — from the ground state to the excited state and back — are examples of induced absorption and induced emission, respectively. Together, these effects create a periodic process known as **Rabi oscillations** with an angular Rabi frequency $\omega_R = 2F/\hbar$.

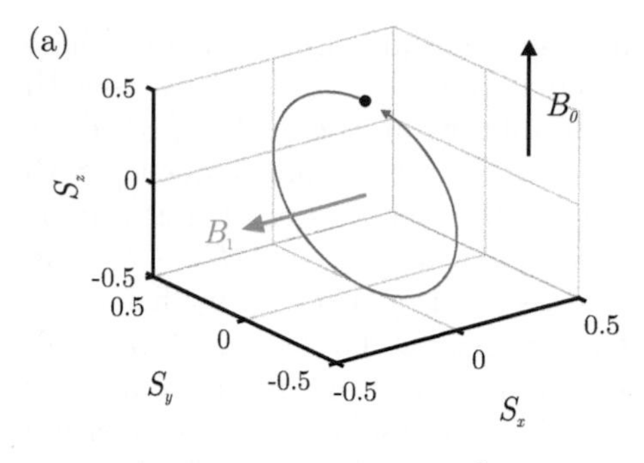

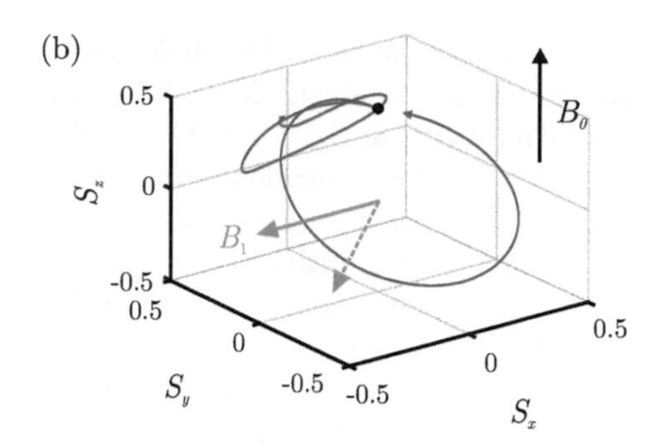

Fig. 10.2 Dynamics of a spin-$\frac{1}{2}$ angular momentum in an external field B_0. Axes are in units of $\hbar$ and the coordinate system rotates around the B_0 axis at an angular frequency $\omega_L = \gamma B_0/\hbar$. The green arrow indicates the direction of a B_1 field orthogonal to B_0. The presence of B_1 leads to Rabi oscillations for an initial state $S_z = \hbar/2$ shown as a black dot. (a) When B_1 rotates at $\omega = \omega_L$ (corresponding to $\Delta_U = 0$), B_1 appears static in the rotating frame and Rabi oscillations occur along full circles. (b) For $\Delta_U \neq 0$, B_1 revolves in the $x-y$ plane, as indicated by a dotted green arrow. In this situation, the Rabi oscillations have a reduced depth, that is, they do not reach the state $S_z = -\hbar/2$. Orange and blue traces indicate the cases $\Delta_U^2 \ll \omega_R^2$ and $\Delta_U^2 \ll \omega_R^2$, respectively.

For $\Delta_U \neq 0$, the imaginary terms in eqns (10.17) and (10.18) cause a change of the phases of the c_i in the complex plane. The equations can then be solved, for arbitrary initial conditions $c_i(0)$, as [138]

$$c_0 = e^{i\Delta_U t/2}\left[c_0(0)\cos\left(\frac{1}{2}\Omega t\right) - \frac{i}{\Omega}\left(\Delta_U c_0 + \omega_R^2 c_1(0)\right)\sin\left(\frac{1}{2}\Omega t\right)\right], \tag{10.19}$$

$$c_1 = e^{i\Delta_U t/2}\left[c_1(0)\cos\left(\frac{1}{2}\Omega t\right) + \frac{i}{\Omega}\left(\Delta_U c_1 + \omega_R^2 c_0(0)\right)\sin\left(\frac{1}{2}\Omega t\right)\right], \tag{10.20}$$

where we introduce the symbol $\Omega \equiv \sqrt{\omega_R^2 + \Delta_U^2}$. The depth of the modulations between the states in the presence of detuning is

$$M_R = \frac{\omega_R^2}{\omega_R^2 + \Delta_U^2}. \tag{10.21}$$

There are two interesting limits to eqn (10.21): when $\Delta_U^2 \gg \omega_R^2$, the Rabi modulation depth decreases approximately as $1/\Delta_U^2$ with the detuning. This is due to the fact that while a Rabi cycle takes place, the relative phase between the drive and the superposition state changes, such that the effect of the force cancels partially. In the opposite limit with $\omega_R^2 \gg \Delta_U^2$, the Rabi oscillation retains a modulation depth of $M_R \approx 1$. The fact that a strong driving force can compensate for the effect of detuning to enable (almost) full inversion cycles in a two-level system is known under the name of **power broadening**. Note that this effect is independent of the state lifetime κ^{-1} but leads to an effective increase in the spectral bandwidth of the system [144], similar to a nonlinear damping term [145]. A detuned force can therefore still elicit state evolutions even when $\Delta_U \gg \kappa$.

We can use the case of a single electron spin in an external field to illustrate these concepts. When a static field B_0 is applied in the z direction, the spin states with angular momentum $S_z = \pm\hbar/2$ (and $Sx = S_y = 0$) differ by an energy $E_{\text{spin}} = \gamma B_0 = \hbar\omega_L$. Here, γ is the so-called *gyromagnetic ratio* and ω_L is the Larmor frequency with which the spin precesses around the magnetic field whenever it is not aligned along its axis. We can identify this as our two-level system with $E_{\text{spin}} = E_1 - E_0$ and $\omega_L = \omega_0 + U$. The role of the external force F is assumed by a secondary field with strenght B_1 that rotates in the $x-y$ plane with frequency ω. For $\omega = \omega_L$ ($\Delta_U = 0$), the drive field is perceived as stationary by the spin in its rotating frame. When the spin is initialized in $|S_z = \hbar/2\rangle$, we observe a precession of the spin angular momentum around B_1 that manifests as Rabi oscillations between $|S_z = \pm\hbar/2\rangle$, see Fig. 10.2(a).

When $\Delta \neq 0$, the drive field B_1 revolves in the $x-y$ plane in the rotating frame. When this rotation is slow relative to the Rabi oscillation, the angle covered by B_1 during a Rabi period $2\pi/\omega_R$ is small. We recognize this as the case $\omega_R^2 \gg \Delta_U^2$ where the Rabi oscillations are only slightly affected, see the orange trace in Fig. 10.2(b). For $\Delta_U^2 \gg \omega_R^2$, the B_1 angle changes rapidly in our rotating frame. Here, no full Rabi oscillations take place and the spin describes small circles close to its initial state, see blue trace in Fig. 10.2(b).

Let us now return to our Kerr resonator and apply the knowledge we gained on pure two-level systems. We found that beyond $|n = 1\rangle$, a strong nonlinearity $U \gg \kappa$ leads to a detuning of the Kerr resonator relative to a force applied at $\Delta_U = 0$. The nonlinearity can thus turn an oscillator into an effective two-level system. In Fig. 10.3, we show that a nonlinear resonator can indeed perform Rabi oscillations by inserting the Hamiltonian in eqn (10.4) into eqn (9.8). The result differs very much from that of the harmonic oscillator in Fig. 9.3: as expected, the system cycles back and forth between the two lowest Fock states, while higher states remain entirely unpopulated, as predicted by eqn (10.16). The Wigner representation in Fig. 9.3(c) confirms that the system can indeed be prepared in the energy eigenstate with $n = 1$, in contrast to the coherent states that were generated in the harmonic oscillator.

When the force is strong relative to the nonlinearity ($F \gg U$), we can invoke power broadening to understand why a force with $\Delta_U = 0$ can overcome the Kerr detuning to populate energy levels beyond $|n = 1\rangle$. In this limit, we can access the semiclassical regime where $\langle n\rangle \gg 1$. Equation (9.8) can then be replaced by eqns (10.11) and (10.12) to study coherent states. Even there, however, the nonlinearity leads to a saturation of the driven amplitude below that of the corresponding harmonic resonator, cf. the amplitude at $\omega = \omega_0$ as a function of $\beta = \beta_3$ in Fig. 2.2(a). This saturation is generated by the same fundamental process that we see at work in Fig. 10.3; the nonlinearity causes a change of the level separations, and thus a shift of the resonance frequency, as a function of the oscillation amplitude. In the classical case, this phenomenon is captured in eqn (2.15).

10.2.2 Quantum System with Parametric Pumping

The next element of the Hamiltonian in eqn (10.4) that we want to investigate is the parametric pumping term with coefficient G, which is proportional to the potential modulation depth λ in the classical model, cf. eqn (10.7). In the quantum picture of

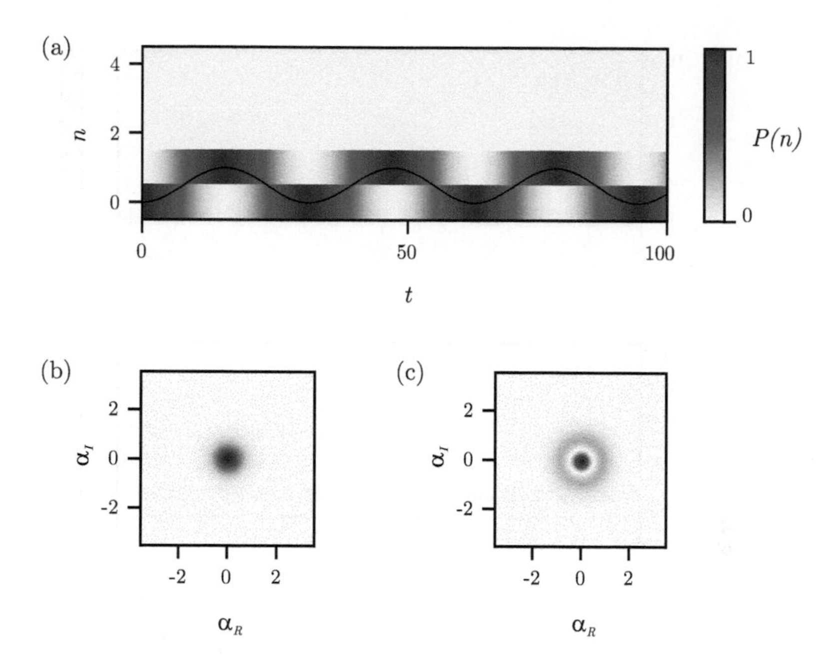

Fig. 10.3 External force applied to an isolated quantum Kerr oscillator initialized in $|0\rangle$, with $m = 1$, $\Delta_U = 0$, $\kappa = 0$, $F = 0.1$, $U = 2$, and $\hbar = 1$. A black line marks $\langle n \rangle$, indicating Rabi oscillations with a frequency $\omega_R/2\pi = F/\hbar\pi = 0.032$. (a) Time evolution of the Fock spectrum. (b) Wigner distribution of the oscillator state at $t = 0$. (c) Wigner distribution of the oscillator state at $t = 78$.

three-wave mixing, this term can be understood as a **two-particle drive**, interpreted as a particle with energy $2\hbar\omega$ that splits into two particles of energy $\hbar\omega \approx \hbar\omega_0$. In a closed system ($\kappa = 0$), a two-particle drive causes a harmonic oscillator ($U = 0$) to leave out every second rung while climbing the Fock ladder: when initialized in $|0\rangle$, the system will only develop finite values of $P(n)$ for the levels $|n = 2j\rangle$ with a natural number j. For an initialization in $|1\rangle$, the levels with finite probabilities will be $|n = 2j + 1\rangle$. This parity conservation can clearly be seen in Fig. 10.4.

For the closed harmonic system shown in Fig. 10.4, there is no mechanism to limit the energy added to the oscillator, so the system will grow without bound. Just like in the classical case discussed in Section 3.1.1, we can avoid this unlimited growth by including dissipation to the environment ($\kappa > 0$). We know that for $\omega = \omega_0$, the instability threshold is reached when $\lambda = 2/Q$. We can translate this to our quantum formalism by inserting $\kappa = \Gamma$ and G from eqn (10.7), and we find the threshold value for G to be at

$$G_{\text{th}} = \frac{\hbar\kappa}{2}. \tag{10.22}$$

In Fig. 10.5, we visualize the combined effect of dissipation and parametric pumping below threshold. The system resides predominantly in the ground state $|n = 0\rangle$, since the two-particle drive is compensated by single-particle loss. However, we observe

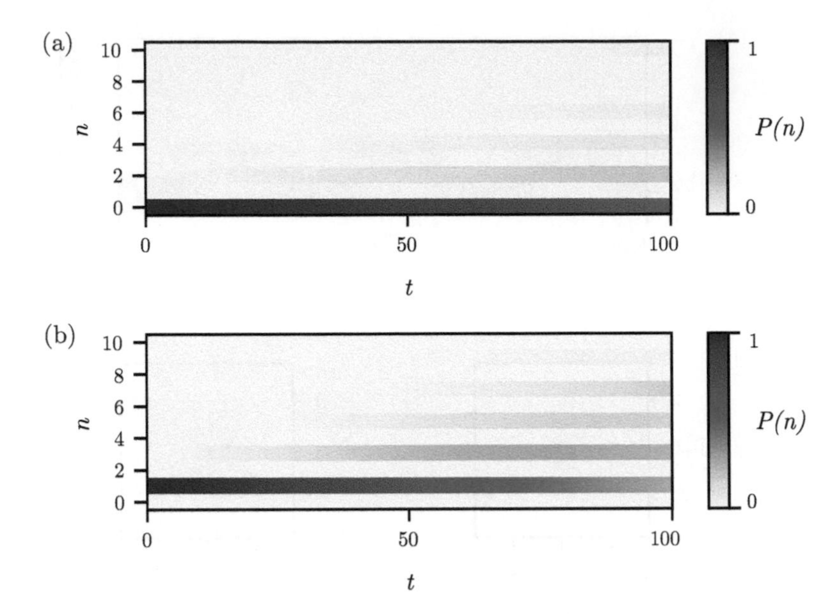

Fig. 10.4 Parametric pumping applied to a quantum oscillator initialized in (a) $|0\rangle$ and (b) $|1\rangle$, with $m = 1$, $\Delta_U = 0$, $\kappa = 0$, $F = 0$, $U = 0$, $G = 0.01$, and $\hbar = 1$.

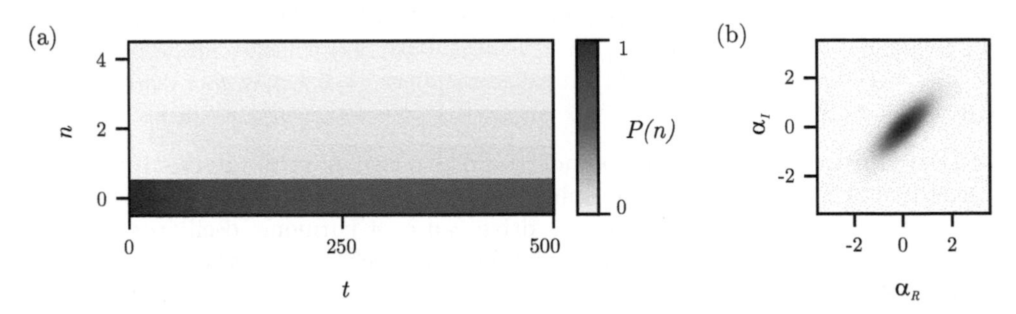

Fig. 10.5 Parametric pumping applied to a quantum oscillator initialized in $|0\rangle$, with $m = 1$, $\Delta_U = 0$, $\kappa = 0.025$, $F = 0$, $U = 0$, $G = 0.01$, and $\hbar = 1$. The environment has a mean excitation number $n_{\text{th}} = 0$. (a) Time evolution of the Fock spectrum. (b) Wigner distribution of the oscillator state at $t = 500$.

squeezing of the quantum ground state. In one (diagonal) quadrature, the uncertainty of the squeezed state is below that of the ground state, while in the opposite quadrature it is increased. This is exactly the same phenomenon as observed for a squeezed thermal state in Chapter 5, except that the cloud in phase space originates from quantum fluctuations, and not from coupling to a thermal environment.

When $G > G_{\text{th}}$, the single-particle loss cannot compensate for the two-particle drive and the energy added can only be limited by the nonlinearity U, similarly to what we have seen in Section 10.2.1. For the example in Fig. 10.5, this point is reached for $G = 0.01$ and $\kappa < 0.02$. From the classical treatment in Chapter 3, it is clear that in the semiclassical limit without detuning, the system should have two attractors (i.e. *phase states*). In the quantum system, each attractor can be approximated as a coherent state $|\tilde{\alpha}\rangle$, and we can use eqn (10.8) to calculate

$$\tilde{\alpha} = \pm\sqrt{G/U} \tag{10.23}$$

for $\hbar\kappa \ll G$ [63, 146]. This result is the quantum notation of eqn (3.13). The coherent states $|\pm\tilde{\alpha}\rangle$ are degenerate eigenstates of the Hamiltonian in eqn (10.4) [83], in analogy with what we obtain for a classical system with eqn (3.4) in the limit $Q^2\lambda^2 \gg 1$.

In a classical parametric oscillator, the system performs a spontaneous breaking of the time-translation symmetry when selecting one of the phase states. On the microscopic level, the symmetry is broken by classical fluctuations that are random in phase, and the term *spontaneous* merely reflects this randomness, cf. Chapter 5. In the quantum system, by contrast, the uncertainty of the system in phase space is dominated by quantum fluctuations. As we discussed in Section 8.1.1, quantum fluctuations are not the consequence of lacking knowledge, but represent the fact that the system is in a superposition between different possible states. This simultaneous realization of different states has fundamental consequences: starting from the quantum ground state, the system does not need to perform a symmetry breaking in response to a parametric drive. Rather, it can ring up to a superposition of *both* attractors simultaneously and can be described in the steady state as [146]

$$\rho = c_+ |C_+\rangle \langle C_+| + c_- |C_-\rangle \langle C_-| \,, \tag{10.24}$$

where the so-called **cat states** $|C_\pm\rangle$ are superpositions of the coherent states,

$$|C_\pm\rangle = c_{\pm\mathrm{cat}} \left(|\tilde{\alpha}\rangle \pm |-\tilde{\alpha}\rangle\right) . \tag{10.25}$$

The normalization factor is $c_{\pm\mathrm{cat}} = 1/\sqrt{2}$ for $\langle n\rangle = \left|\tilde{\alpha}^2\right| \to \infty$. For small $\tilde{\alpha}$, the two coherent states in a cat state overlap partially,

$$|\langle\tilde{\alpha}| - \tilde{\alpha}\rangle| = e^{-2\langle n\rangle} \,, \tag{10.26}$$

and the normalization becomes [63, 83]

$$c_{\pm\mathrm{cat}} = 1/\sqrt{2\left(1 \pm e^{-2\langle n\rangle}\right)} \,. \tag{10.27}$$

In a closed system, the parametric pump term (two-particle drive) conserves the parity of the Fock spectrum, cf. Fig. 10.4. The resulting states are highly non-classical and include only odd or even Fock states when initialized in $|0\rangle$ or $|1\rangle$, respectively. In Fig. 10.6, by contrast, we show the evolution of a parametric oscillator driven above threshold G_{th} in the presence of a significant single-particle loss κ to the environment. Single-particle loss breaks the parity on a timescale κ^{-1}, such that the probability densities of the states retain no negativity and resemble the classical counterpart in Fig. 5.9 at first glance. In the quantum system, however, ρ does not represent the statistical distribution that emerges over long timescales, but the superposition state that the system realizes at a single point in time.

We can restore the parity conservation of the Kerr parametric oscillator above threshold when κ becomes sufficiently small. We first note that for $G \to 0$, the cat

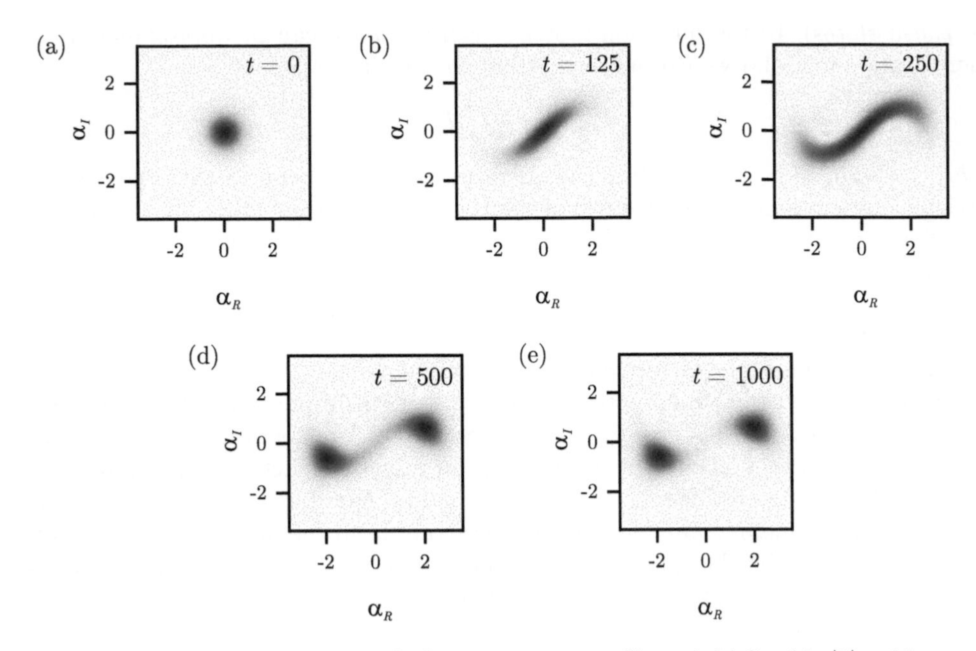

Fig. 10.6 Parametric pumping applied to a quantum oscillator initialized in $|0\rangle$, with $m = 1$, $\Delta_U = 0$, $\kappa = 0.01$, $F = 0$, $U = 0.002$, $G = 0.01$, and $\hbar = 1$. The environment has a mean excitation number $n_{\text{th}} = 0$. The individual panels show the Wigner quasiprobability distribution at different times t.

states smoothly transition into Fock states, namely $|C_+\rangle \to |0\rangle$ and $|C_-\rangle \to |1\rangle$. According to the **adiabatic theorem**, it should therefore be possible to initialize the system in one of the Fock states and adiabatically evolve it into the corresponding cat state [147]. Here, *adiabatic* means that we do not switch on the parametric pump suddenly, but ramp it from $G = 0$ to the desired value on a timescale that is slow compared to $\hbar/U$. At the same time, the evolution must be much faster than the timescale κ^{-1} over which single-particle loss breaks the parity. A simulation showing such a cat state generation is shown in Fig. 10.7. Experimentally, creating a situation where the required condition $U \gg \hbar\kappa$ is fulfilled is very challenging [63].

10.2.3 Parametric Pumping and External Force

How does the system respond when a weak external force F is applied in addition to the parametric pump? The external force can be identified as a *single-particle drive* which can break the parity of the Fock spectrum, in contrast to the two-particle drive G. Differently than the single-particle loss mechanism with rate κ, the force acts coherently and equally on all Fock states to lift the degeneracy between the eigenstates $|\pm\tilde{\alpha}\rangle$. As a result, the system performs Rabi oscillations with an angular frequency $\omega_R = Re(4F\tilde{\alpha})$ between the cat states $|C_+\rangle$ and $|C_-\rangle$ under the influence of F [63].

In Fig. 10.8, we can observe such Rabi oscillations. Note that the energy expectation value $\langle n\rangle$ does not shift noticeably with time for such a small force, and that the symmetry between the left and right halves of the phase space representation is

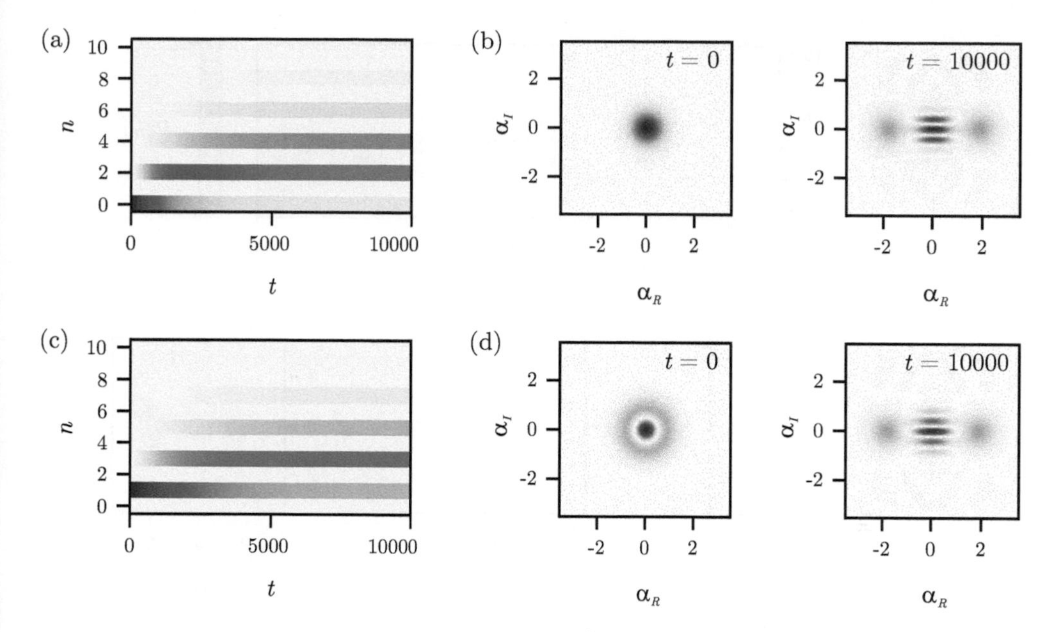

Fig. 10.7 Parametric pumping $G = 0.4 \times (1 - e^{-t/t_G})$ applied to a quantum oscillator with $m = 1$, $\Delta_U = 0$, $\kappa = 0$, $F = 0$, $U = 0.12$, $t_G = 2000$, and $\hbar = 1$. The environment has a mean excitation number $n_{\text{th}} = 0$. (a) Fock spectrum for an oscillator initialized in $|0\rangle$, with the same color bar for $P(n)$ as in Fig. 10.5. (b) Wigner representations of the initial state $|0\rangle$ and the final state $|C_+\rangle$ of the evolution shown in (a). (c)-(d): corresponding graphs for a system initialized in $|1\rangle$ and reaching $|C_-\rangle$.

conserved to a good approximation. By contrast, the symmetry is visibly broken when the force is increased, see Fig. 10.9. This is the quantum version of the parametric symmetry breaking that we introduced for a classical system in Section 3.1.4.

10.3 Coupled Quantum Parametric Oscillators

To conclude this chapter, we provide an outlook onto systems of coupled quantum parametric oscillators. Classical parametric oscillator networks were explored in Chapter 7. The physics of the corresponding quantum systems is being actively researched, and we consider it premature to attempt a meaningful summary at this point. However, we can combine previous knowledge to make simple predictions about the long-time behavior of resonator networks in the semiclassical regime ($n \gg 1$).

We formulate a quantum version of eqn (7.1) starting from the RWA Hamiltonian in eqn (10.3) for $F = 0$. The coupling terms are implemented as mutual driving terms that annihilate a particle on resonator j to create one on resonator k or vice versa, resulting in the Hamiltonian [146]

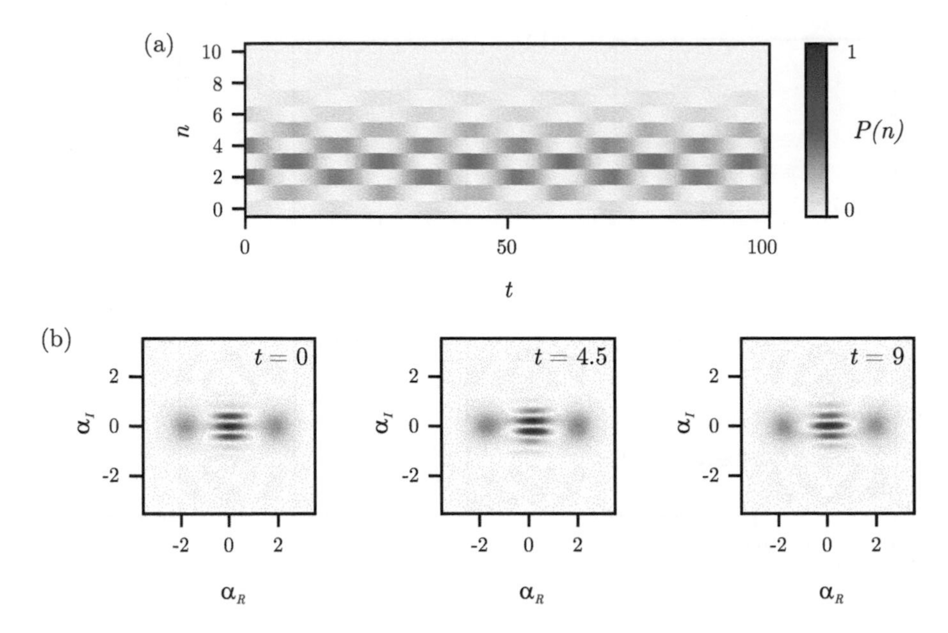

Fig. 10.8 Parametric pumping $G = 0.4$ and external force $F = 0.05$ applied simultaneously to a quantum oscillator with $m = 1$, $\Delta_U = 0$, $\kappa = 0$, $U = 0.12$, $\theta = 0$, and $\hbar = 1$. The environment has a mean excitation number $n_{\text{th}} = 0$. (a) Fock spectrum for an oscillator initialized in $|C_+\rangle$. (b) Wigner representations of the states at various times, corresponding to even, mixed, and odd cat states from left to right, respectively.

$$\tilde{H} = \sum_j \left[-\hbar\Delta_{U,j}\tilde{a}_j^\dagger\tilde{a}_j + \frac{U_j}{2}\tilde{a}_j^{\dagger 2}\tilde{a}_j^2 - \frac{G_j}{2}\left(\tilde{a}_j^{\dagger 2} + \tilde{a}_j^2\right) + \sum_{k \neq j} \frac{\hbar J_{jk}}{2}(\tilde{a}_j^\dagger\tilde{a}_k + \tilde{a}_j\tilde{a}_k^\dagger) \right] . \tag{10.28}$$

As in the classical system, coupled resonators generally form normal modes. Here, the nonlinear terms appear both as self-Kerr terms, i.e., nonlinear potential terms for each normal mode, and cross-Kerr terms, which restore coupling between the normal modes [107]. We have already encountered these terms in eqn (7.13). For weak nonlinearities, we ignore the latter and assume that our network can be approximated as uncoupled normal modes. Similarly, the two-photon drives G_j generate both normal-mode squeezing and two-mode squeezing terms, of which the former give rise to parametric instability. Neglecting the influence of two-mode squeezing, we then obtain a normal-mode Hamiltonian

$$\tilde{H} = \sum_j \left[-\hbar\Delta_{U,j}\tilde{a}_j^\dagger\tilde{a}_j + \frac{U_j}{2}\tilde{a}_j^{\dagger 2}\tilde{a}_j^2 - \frac{G_j}{2}\left(\tilde{a}_j^{\dagger 2} + \tilde{a}_j^2\right) \right] , \tag{10.29}$$

where the index j now indicates different normal modes instead of different resonators.

To include the effects of damping and environment fluctuations, we insert the normal-mode Hamiltonian in eqn (10.29) into a Lindblad master equation, cf. eqn (9.8).

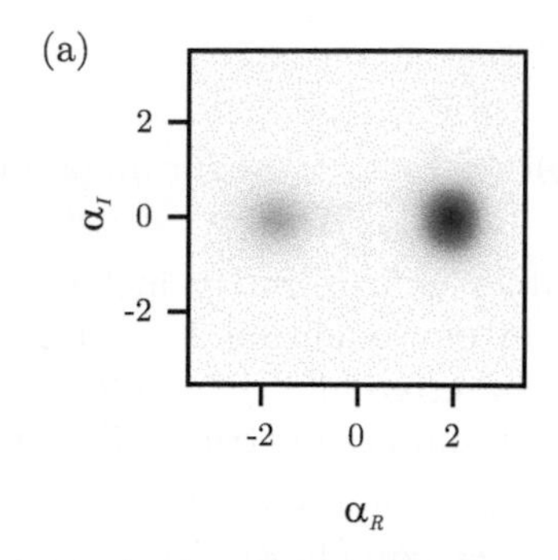

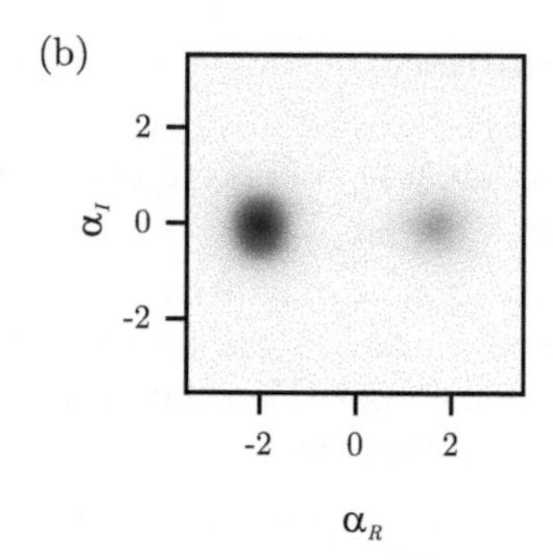

Fig. 10.9 Parametric pumping $G = 0.4$ and external force $F = 0.1$ applied simultaneously to a quantum oscillator with $m = 1$, $\Delta_U = 0$, $\kappa = 0.01$, $U = 0.12$, and $\hbar = 1$. The environment has a mean excitation number $n_{\mathrm{th}} = 0$. (a) and (b) show the Wigner representations of the steady states for $\theta = 0$ and $\theta = \pi$, respectively.

As long as we are only interested in the expectation values of the amplitudes in the semiclassical limit, we can follow the same procedure as in Section 10.1.1 to obtain, in this case, slow-flow equations of the same form as eqn (7.9). We arrive at the conclusion that we can understand the general properties of the quantum network in the long-time, many-particle limit from the classical physics we studied in Chapter 7. However, going to the few-particle limit (with strong coupling) below the decoherence time, we expect to find a host of new phenomena related to the quantum superposition of nonlinear, coupled states and two-mode squeezing.

Chapter summary

- In Chapter 10, we combine all our prior knowledge to study quantum parametric oscillators.
- Our preparation in Chapters 8 and (9) enables us to directly include a Kerr nonlinearity U and a parametric pump G in our quantum Hamiltonian, cf. eqns (10.3) and 10.4. This Hamiltonian can be inserted into the Lindblad master equation to obtain the time evolution of a quantum parametric oscillator.
- Similar to what was done in Section 9.2.3, we can simplify the Lindblad master equation to obtain an equation of motion for $\tilde{\alpha}$, cf. eqn (10.8). With appropriate approximations for a semiclassical (large-n) limit, this equation of motion yields the classical slow-flow equations, cf. eqns (10.11) and (10.12).
- In order to build up an understanding of the full **quantum Kerr parametric oscillator**, we first study the quantum Kerr oscillator ($G = 0$). We find that the nonlinear potential term can lead to an effective two-level system and to **Rabi oscillations** under an external drive, cf. Fig. 10.3.
- With $G > 0$, we recover many familiar features encountered in the classical stochastic system in Chapter 5, such as squeezing and the ringup to phase states beyond a parametric threshold, cf. Figs. 10.5 and 10.6. In contrast to the classical system, however, the quantum system can be prepared in a **quantum superposition of coherent states**, and in states comprising only even or odd Fock states. As the parametric pump corresponds to a two-particle drive and conserves parity, the system reaches an even **cat state** $|C_+\rangle$ when starting from $|0\rangle$ and an odd cat state $|C_-\rangle$ when starting from $|1\rangle$, cf. Fig. 10.7. A weak external drive can be used to toggle between parity states, leading to Rabi oscillations between even and odd cat states, cf. Fig. 10.8.
- We finish with a succinct look at networks of quantum Kerr parametric oscillators in the semiclassical limit.

Exercises

Check questions:

(a) When applying a semiclassical equation of motion, what features are we potentially missing? In what case can the result for $\tilde{\alpha}$ be wrong?

(b) Why is the two-level system sometimes invoked as the opposite of a harmonic oscillator?

(c) What is the effect when a strong Kerr nonlinearity U is added to a quantum harmonic oscillator? Why does the response of the system to an external force depend on the size of F_0?

(d) Can you guess what the Fourier spectrum of a quantum Kerr oscillator residing in a coherent state would look like for $\kappa = 0$?

(e) What happens to a harmonic quantum oscillator ($U = \kappa = 0$) when it is driven parametrically? What are the differences between systems with or without loss κ?

(f) Why did we not discuss the parity of a parametrically pumped system in the classical limit?

(g) Discuss the criteria for observing cat states experimentally.

(h) Which of the quantum phenomena generated by a symmetry-breaking force in Section 10.2.3 have a classical analogue? Which ones do not, and why?

Tasks:

10.1 In **Python Example 9**, use $N = 22$, $\Delta_U = 0$, $G = \kappa = 0$, $U/\hbar = 2$, and $n_{\mathrm{th}} = 0$. Initialize the oscillator in $|n = 0\rangle$ and drive it with various forces. Check if the resulting dynamics follows Rabi oscillations with the correct frequency.

10.2 Detune the driving frequency ($\Delta_U \neq 0$) and test if you can observe the effect power broadening for various force strengths. What is a good metric to quantify the effect? Use eqns (10.17) and (10.18) to reproduce your results.

10.3 Initialize your system in $|n = 0\rangle$ and use a pulsed (time-dependent) F to drive the oscillator into an equal superposition of $|n = 0\rangle$ and $|n = 1\rangle$, leave it there for a certain amount of time, and then drive it back to $|n = 0\rangle$ with a second pulse. What changes when you add finite dissipation κ and a thermal occupation of the bath? How can you (qualitatively) define from this numerical experiment a decoherence time?

10.4 Can you change the code in such a way that, starting from $|n = 0\rangle$, you can drive the system into the state $|n = 2\rangle$ or $|n = 3\rangle$?

10.5 Simulate the effect of parametric amplification below threshold in a quantum system. Use $U = 0$ for simplicity. Can you explain all the features you see?

10.6 Set $F = 0$ and $U/\hbar = 0.1$ and use initialization in $|n = 0\rangle$ or $|n = 1\rangle$ to parametrically drive the system into even or odd cat states, respectively. Use an appropriate value of $\mathbf{tG}$ to engineer an adiabatic evolution. Test the influence of loss κ or finite temperature on the coherence of the process.

10.7 Repeat the simulation starting from the superpositions $\frac{1}{\sqrt{2}}(|n = 0\rangle + |n = 1\rangle)$ and $\frac{1}{\sqrt{2}}(|n = 0\rangle + i\,|n = 1\rangle)$. How do the Fock-state distributions of these two states compare to each other? How do they compare to the one of the even and odd cat states?

10.8 Once the system is in a cat state, use a small force F to switch between even and odd cat states. Test if the dynamics follows the Rabi frequency $\omega_R = \mathrm{Re}\,[4F\tilde{\alpha}]$.

10.9 Simulate the steady-state amplitude of the quantum parametric oscillator as a function of detuning and compare the result to classical and quantum models, cf. eqns (3.4) and (10.8). In what aspects do the classical and quantum models differ?

11
Experimental Systems

In Chapters 1 to 10, we progressed from the model of a single, deterministic, harmonic oscillator to the complex stochastic behavior of coupled, nonlinear quantum resonators. In all of these treatments, we used as an example a point-like particle in a potential well, where the meaning of parameters like mass, spring constant, or nonlinearity is immediately clear. Many other physical systems can be mapped onto the same model, but the translation of an initial physical description into an EOM can be less straightforward. We also limited our previous discussion to dimensionless parameters with a convenient range of values, for example $\omega_0 = 1$, instead of realistic and dimensionfull values such as $\omega_0/2\pi = 2.5 \times 10^8$ Hz.

Below, we present several realistic examples of resonators. For each of them, we derive the corresponding EOM to allow for direct applications of the equations in the previous chapters. Furthermore, we discuss how parameter values can be scaled to avoid computational problems with numerical simulations.

Note that we reuse some symbols in order to adhere to the notation employed in various communities. For instance, L will be used for a length in Section 11.1 and for an inductance in Section 11.2. Likewise, E was generally understood to stand for an energy in the previous chapters, but will denote Young's modulus in Section 11.1 and an electric field in Section 11.3.

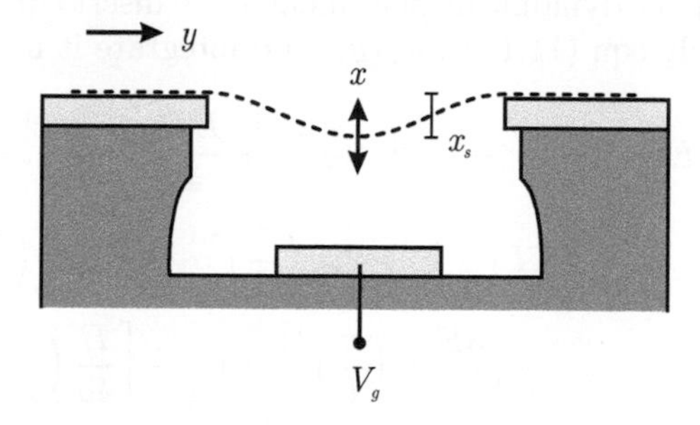

Fig. 11.1 Sketch of a carbon nanotube (dashed line) clamped on two metal electrodes (light gray) and suspended like a string over a trench with a third *gate* electrode. A gate voltage V_g can be applied to the third electrode to generate static and time-dependent forces, resulting in a static deflection x_s and an oscillating deflection x, respectively.

11.1 Mechanical Resonator Example

Our first example is a mechanical resonator made from a single carbon nanotube, see Fig. 11.1. A nanotube is a tubular molecule consisting of a sheet of carbon atoms arranged in a honeycomb pattern [148, 149]. In our example, we consider a tube with length L that is clamped at both ends. Following Ref. [150], we apply the Euler–Bernoulli equation to this beam and obtain

$$\rho S \frac{\partial^2 x_b}{\partial t^2} = -EI \frac{\partial^4 x_b}{\partial y^4} + \left[T_{in} + \frac{ES}{2L} \int_0^L \left(\frac{\partial x_b}{\partial y} \right)^2 \partial y \right] \frac{\partial^2 x_b}{\partial y^2} + f_b(t) \, . \tag{11.1}$$

In this equation, t is time and $x_b = x_b(y,t)$ is the beam displacement along the axis coordinate y. We use ρ to designate the nanotube's mass density, E for Young's modulus, S for the cross-sectional area, I for the second moment of inertia about the longitudinal axis, T_{in} for the built-in tension, and $f_b(t)$ for a force per unit length along the beam, which can be created by applying a voltage to a gate electrode below the nanotube [148].

We assume that the nanotube has low enough bending rigidity to behave like a guitar string. The shape of the lowest mode can then be approximated by the normalized profile along the beam

$$\phi_p(y) = \sin(\pi y / L) \, . \tag{11.2}$$

This profile can be displaced both statically and dynamically,

$$x_b(y,t) = x_s \phi_p(y) + x(t) \phi_p(y) \, , \tag{11.3}$$

where x_s is the maximum static displacement and $x(t)$ is the maximum dynamic displacement along the profile at any given time t. Both displacements are in the plane perpendicular to the gate electrode.

To find an EOM for the dynamic displacement x, we insert eqn (11.2) and eqn (11.3) into eqn (11.1), multiply eqn (11.1) by $\phi_p(y)$, and integrate it from 0 to L to get [150]

$$\begin{aligned} \frac{\partial^2 x(t)}{\partial t^2} = & -\frac{1}{\rho S} \left[EI x_s \left(\frac{\pi}{L} \right)^4 + T_{in} x_s \left(\frac{\pi}{L} \right)^2 + \frac{ES}{4} x_s^3 \left(\frac{\pi}{L} \right)^4 - \frac{4}{\pi} f_b(t) \right] \\ & - \frac{1}{\rho S} \left[EI \left(\frac{\pi}{L} \right)^4 + T_{in} \left(\frac{\pi}{L} \right)^2 + \frac{3}{4} ES x_s^2 \left(\frac{\pi}{L} \right)^4 \right] x(t) \\ & - \left[\frac{3E}{4\rho} x_s \left(\frac{\pi}{L} \right)^4 \right] x^2(t) - \left[\frac{E}{4\rho} \left(\frac{\pi}{L} \right)^4 \right] x^3(t) \, . \end{aligned} \tag{11.4}$$

In general, we may divide the force into a static and a time-dependent component, $f_b(t) = f_s + f(t)$. When the beam is in static equilibrium with f_s, the static terms in the first bracket on the right-hand side of eqn (11.4) must cancel out,

$$EI x_s \left(\frac{\pi}{L} \right)^4 + T_{in} x_s \left(\frac{\pi}{L} \right)^2 + \frac{ES}{4} x_s^3 \left(\frac{\pi}{L} \right)^4 = \frac{4}{\pi} f_s \, . \tag{11.5}$$

The remaining terms of eqn (11.4) can then be reordered to obtain the usual form of a nonlinear EOM without damping,

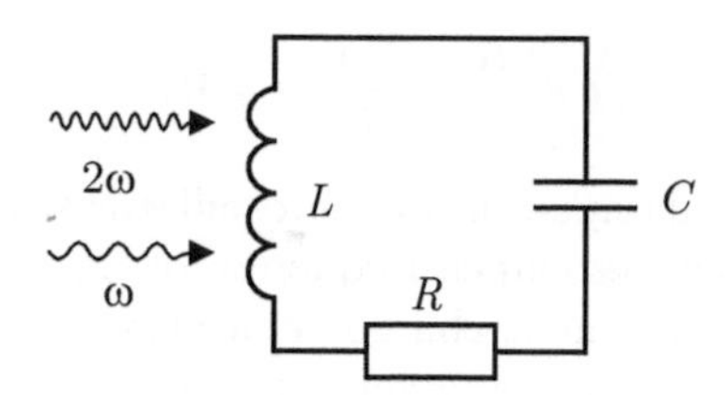

Fig. 11.2 Sketch of an electrical circuit comprising a resistor R, an inductor L and a capacitor C. Driving signals, such as direct forcing at ω and parametric pumping at 2ω, can be applied inductively.

$$\ddot{x}(t) + \omega_0^2 x(t) + \beta_2 x^2(t) + \beta_3 x^3(t) = \frac{F_0}{m}\cos(\omega t)\,, \tag{11.6}$$

where we assume that the time-dependent force corresponds to a single drive tone, $\frac{4}{\pi\rho S} f(t) = \frac{F_0}{m}\cos(\omega t)$. To understand the origin of mechanical nonlinearities, it is interesting to compare eqn (11.4) to eqn (11.6). We find that both β_2 and β_3 are always positive, and that $\beta_2 \propto x_s$, meaning β_2 will disappear for a straight nanotube. It is the broken static symmetry that gives rise to a cubic potential as introduced in Section 2.2 [151].

In order to add parametric driving to eqn (11.6), we simply consider a second drive tone close to $2\omega_0$ as in Section 3.2. As was discussed there, the combination of an external force and the quadratic nonlinearity β_2 creates a parametric modulation of the spring constant, cf. eqn (3.31) [27]. Linear and nonlinear damping terms can often not be derived microscopically for nanomechanical systems and are therefore introduced phenomenologically to fit experimental observations [145]. With all of these terms, we arrive at an EOM of the desired form with an effective nonlinearity β calculated according to eqn (2.35),

$$\ddot{x} + \omega_0^2\left[1 - \lambda\cos\left(2\omega t + \psi\right)\right] x + \beta x^3(t) + \Gamma\dot{x} + \eta x^2\dot{x} = \frac{F_0}{m}\cos(\omega t)\,. \tag{11.7}$$

In the last step, we evaluate the numerical values of the parameters in eqn (11.7). For a typical nanotube with a length of $L = 1.8\,\mu\mathrm{m}$, a radius of $r = 1.5\,\mathrm{nm}$, and a built-in tension of $T_{in} = 0.1\,\mathrm{nN}$, we find values around $m \approx 1\times10^{-20}\,\mathrm{kg}$, $\omega_0 \approx 2\pi\times 50\,\mathrm{MHz}$, and $\beta \approx 3\times10^{31}\,\mathrm{m}^{-2}\,\mathrm{s}^{-2}$ [150]. The linear quality factor Q is in the order of a few hundreds at room temperature to a few thousands at cryogenic temperatures, and up to a few millions for optimized conditions [152]. The nonlinear damping coefficient was estimated to be roughly $\eta \approx 8\times10^{25}\,\mathrm{m}^{-2}\,\mathrm{s}^{-1}$ for a similar device [145].

11.2 Electrical Resonator Example

As a second example of a parametric resonator, we will study a nonlinear electrical circuit. We start from the setup shown in Fig. 11.2, which consists of a resistance R, an inductance L, and a capacitance C connected in a loop. Applying Kirchhoff's second law, we find an equation for the electric potential drop over the various elements in terms of q, which is the charge separated over the capacitor:

$$\ddot{q} + \frac{R}{L}\dot{q} + \frac{1}{LC}q = 0\,. \tag{11.8}$$

Equation (11.8) describes a damped harmonic oscillator with a mass of 1, a damping rate of R/L, and an angular resonance frequency of $\omega_0^2 = 1/LC$ (for $Q \gg 1$). The system oscillates between maximum charge separation reached at $\dot{q} = 0$ (when no current flows through the inductor) and zero charge separation at $q = 0$ (with maximum current flowing). The resistor stores no energy but provides a damping element that, in the absence of a driving force, leads to a ringdown of the oscillation. When no resistor is added to a circuit, the resonator's damping will be defined by the finite conductivity of the wires themselves. This residual resistance can be reduced (almost) to zero by using a superconductor for the circuit, leaving only material defects to limit the quality factor [153].

Adding a nonlinearity to the circuit can be achieved in various ways. In a superconducting circuit, a Josephson junction imparts a nonlinear inductance. Parametric oscillators made from superconducting circuits with Josephson junctions are a powerful resource for quantum devices because they combine (i) GHz frequencies that lead to low thermal occupation in a dilution refrigerator, (ii) large Q factors that enable long coherence times, and (iii) very strong nonlinearities for controlled operation at small particle numbers [52, 63, 154–156].

An alternative route for building a nonlinear RLC circuit that also works at room temperature is to replace the linear capacitor with a nonlinear capacitive element, known as a varactor (or varicap) diode [75]. While a linear capacitor has a potential drop of $U_C = q/C$ as a function of separated charge q, the potential drop across the varactor diode can be approximated as $U_{\text{var}} = aq + bq^2$. With this nonlinear term, eqn (11.8) transforms into

$$\ddot{q} + \omega_0^2 q + \beta_2 q^2 + \Gamma\dot{q} = 0\,, \tag{11.9}$$

with $\omega_0^2 = a/L$, $\beta_2 = b/L$, and $\Gamma = R/L$.

Driving can be provided by voltage signals added for example through inductive coupling or with an auxiliary coil, cf. Fig. 11.2 [75]. Voltage signals with a frequency ω close to ω_0 provide external forcing, while signals at 2ω can be used for parametric pumping in the presence of the quadratic nonlinearity β_2, cf. eqn (3.31). The latter also gives rise to an effective Duffing nonlinearity β, cf. eqn (2.35). Finally, we can introduce a nonlinear damping coefficient η. In Ref. [75], this nonlinear damping term was attributed to the behavior of the varactor diode when the amplitude inside the circuit approaches the forward bias required to open the diode, which leads to a sudden increase in the damping.

Using q as a resonator displacement has the disadvantage that this quantity is not directly measurable. Instead, it is much more convenient to state the EOM in terms of the voltages applied to, and measured, with a lock-in amplifier [75]. Using the measured voltage instead of the charge as the displacement x requires a unit transformation with calibration factors that can be determined experimentally. We then obtain the EOM in the convenient form,

$$\ddot{x} + \omega_0^2\left[1 - \lambda\cos\left(2\omega t + \psi\right)\right]x + \beta x^3(t) + \Gamma\dot{x} + \eta x^2\dot{x} = F_d\cos(\omega t)\,, \tag{11.10}$$

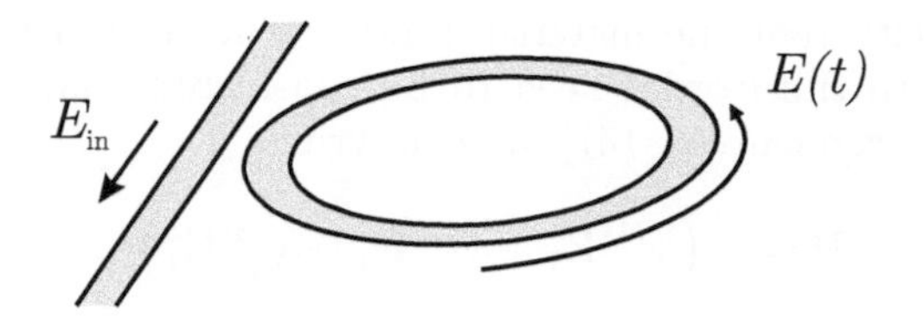

Fig. 11.3 Sketch of an optical ring resonator coupled weakly to an optical waveguide for input and output signals. Light gray marks the optical material.

where $F_d = cU_d/L$ with U_d the drive voltage applied to the auxiliary coil and c the corresponding calibration factor.

The numerical values found in Ref. [75] for such a device are $\omega_0 \approx 2\pi \times 3.4\,\text{MHz}$, $\beta \approx 3 \times 10^{17}\,\text{V}^{-2}$, $Q \approx 240$, $F_d/U_d \approx 2.7 \times 10^{10}\,\text{s}^{-2}$, and $\eta \approx 6 \times 10^{8}\,\text{V}^{-2}\,\text{s}^{-1}$. The parametric modulation depth is equal to $\lambda = 2U_p/(U_{th}Q)$, where U_p is the voltage amplitude applied at 2ω and $U_{th} \approx 4.8\,\text{V}$ is the threshold voltage above which parametric oscillations appear.

11.3 Optical Resonator Example

In our third example, we consider an optical ring resonator, see Fig. 11.3. The electric field within the waveguide is governed by the nonlinear wave equation

$$\nabla^2 \mathbf{E} - \frac{1}{c^2}\frac{\partial^2 \mathbf{E}}{\partial t^2} = \mu_0 \frac{\partial^2 \mathbf{P}}{\partial t^2}\,, \tag{11.11}$$

where $\mathbf{E} = \mathbf{E}(\mathbf{r}, t)$ is the electric field strength at a position $\mathbf{r}$ and time t, c is the speed of light, μ_0 is the vacuum permeability, and $\mathbf{P}$ is the polarization density of the optical material.

The electric field inside the ring forms standing waves when the circumference of the ring L is an integer multiple of the wavelength λ_{opt}. Each integer ratio L/λ_{opt} defines an optical resonant mode with a normalized spatial vector field profile $\phi_i(\mathbf{r})$ and a time-dependent electric strength $E_i(t)$ that we assume to be polarized in the radial direction $\boldsymbol{e}_{\text{rad}}$ relative to the ring structure,[1]

$$\mathbf{E}_i(\mathbf{r}, t) = E_i(t)\phi_i(\mathbf{r})\boldsymbol{e}_{\text{rad}}\,, \tag{11.12}$$

such that

$$\mathbf{E} = \sum_i \mathbf{E}_i\,. \tag{11.13}$$

Here, the field strength E_i represents the displacement of the standing wave. In the following, we consider a single resonant mode with index 1 and ignore the additional complexity that will in reality arise from coupling to other modes.

The polarization density inside the optical ring can have nonlinear components of order n quantified by the tensors $\chi^{(n)}$, which arise from the material properties.

[1] Assuming a different polarization does not influence the following results for a single mode, but adds an additional degree of freedom for obtaining different modes inside the waveguide.

Assuming for simplicity that the material is homogeneous and isotropic,[2] that only nonlinear effects up to third order need to be considered, and that dispersion and phase matching are taken care of [157], we can write

$$\mathbf{P} = \epsilon_0 \left(\chi^{(1)} \mathbf{E}_1 + \chi^{(2)} \mathbf{E}_1^2 + \chi^{(3)} \mathbf{E}_1^3 \right) . \tag{11.14}$$

Note that for simplicity we include only a single mode R_1 in the polarization expansion. Generally, the full mode expansion from eqn (11.13) is used in eqn (11.14), leading to inter-mode hybridization terms that are commonly treated as additional sources for linear and nonlinear damping. As we are not interested in the spatial distribution of the mode, we integrate over the entire volume V to obtain an equation for E_1

$$\begin{aligned} E_1 c^2 \int_V \phi_1 \nabla^2 \phi_1 dV' &= \frac{\partial^2 E_1}{\partial t^2} \int_V \phi_1^2 dV' \left(1 + \chi^{(1)} \right) \\ &+ \chi^{(2)} \frac{\partial^2 E_1^2}{\partial t^2} \int_V \phi_1^3 dV' + \chi^{(3)} \frac{\partial^2 E_1^3}{\partial t^2} \int_V \phi_1^4 dV' , \end{aligned} \tag{11.15}$$

where dV' is an integration variable and we ignore terms of quadratic order in $\chi^{(1)}$. Integrating the first term in eqn (11.15) by parts, we can rewrite the equation as

$$\begin{aligned} \ddot{E}_1 \left[1 + \chi^{(1)} + 3\chi^{(3)} E_1^2 \frac{I_4}{I_2} + 2\chi^{(2)} E_1 \frac{I_3}{I_2} \right] + 2\dot{E}_1^2 \chi^{(2)} \frac{I_3}{I_2} \\ + E_1 \left[c^2 \frac{I_1}{I_2} + 6\chi^{(3)} \dot{E}_1^2 \frac{I_4}{I_2} \right] = 0 , \end{aligned} \tag{11.16}$$

where we use the notations

$$I_1 = \int_V \left(\nabla \phi_1 \right)^2 dV' , \tag{11.17}$$

$$I_2 = \int_V \phi_1^2 dV' , \tag{11.18}$$

$$I_3 = \int_V \phi_1^3 dV' , \tag{11.19}$$

$$I_4 = \int_V \phi_1^4 dV' . \tag{11.20}$$

In order to obtain an equation of motion in a familiar form, we divide eqn (11.16) by the first square bracket, yielding

$$\ddot{E}_1 + \frac{2\dot{E}_1^2 \chi^{(2)} \frac{I_3}{I_2} + E_1 \left[c^2 \frac{I_1}{I_2} + 6\chi^{(3)} \dot{E}_1^2 \frac{I_4}{I_2} \right]}{1 + \chi^{(1)} + 3\chi^{(3)} E_1^2 \frac{I_4}{I_2} + 2\chi^{(2)} E_1 \frac{I_3}{I_2}} = 0 . \tag{11.21}$$

We then expand the right-hand side of eqn (11.21) to lowest order and neglect terms of fourth and higher order in E_1, resulting in

[2] In reality, the $\chi^{(2)}$ nonlinearity cannot be isotropic because it arises from a broken symmetry in the crystal. The effective nonlinearity will therefore have a reduced average value.

$$\ddot{E}_1 + E_1 \left[\frac{\omega_0^2}{n_0} + \dot{E}_1^2 \left(\frac{6 I_4 \chi^{(3)}}{I_2 n_0} + \frac{2 I_3 \chi^{(2)}}{I_2 n_0} - \frac{4 I_3^2 \left(\chi^{(2)}\right)^2}{I_2^2 n_0^2} \right) \right]$$
$$- E_1^2 \frac{2 I_3 \chi^{(2)} \omega_0^2}{I_2 n_0^2} + E_1^3 \omega_0^2 \frac{4 I_3^2 \left(\chi^{(2)}\right)^2 - 3 I_2 I_4 n_0 \chi^{(3)}}{I_2^2 n_0^3} = 0\,, \tag{11.22}$$

using the notations $n_0 = 1 + \chi^{(1)}$ and $\omega_0^2 = c^2 \frac{I_1}{I_2}$ for convenience. We furthermore assume that our optical resonator is overcoupled, meaning that damping and external driving are both dominated by the coupling to a secondary waveguide, cf. Fig. 11.3. Without going into the details of how this coupling can be calculated realistically [158], we include these terms phenomenologically into eqn (11.16) and arrive at

$$\ddot{E}_1 + E_1 \left[\frac{\omega_0^2}{n_0} + \dot{E}_1^2 \left(\frac{6 I_4 \chi^{(3)}}{I_2 n_0} + \frac{2 I_3 \chi^{(2)}}{I_2 n_0} - \frac{4 I_3^2 \left(\chi^{(2)}\right)^2}{I_2^2 n_0^2} \right) \right] - E_1^2 \frac{2 I_3 \chi^{(2)} \omega_0^2}{I_2 n_0^2}$$
$$+ E_1^3 \omega_0^2 \frac{4 I_3^2 \left(\chi^{(2)}\right)^2 - 3 I_2 I_4 n_0 \chi^{(3)}}{I_2^2 n_0^3} + \Gamma \dot{E}_1 = F_E \cos\left(2\omega t + \psi\right). \tag{11.23}$$

Next, we employ the averaging method to obtain slow-flow equations for eqn (11.23). We use second-order averaging for the terms containing $\chi^{(2)}$ to preserve the broken potential symmetry, which we need in the following for parametric pumping induced by the drive F_E at the frequency $2\omega \approx 2\omega_0$, cf. eqn (3.31). The resulting slow-flow equations in terms of averaged quadratures $E_1 = u(t)\cos(\omega t) - v(t)\sin(\omega t)$ are

$$\dot{u} = -u\frac{\Gamma}{2} + F_E C_1 \left(u \sin\psi - v\cos\psi\right) - v\left(\left(u^2 + v^2\right) C_2 - \frac{\omega^2 - \omega_0^2}{2\omega}\right), \tag{11.24}$$

$$\dot{v} = -v\frac{\Gamma}{2} - F_E C_1 \left(u \cos\psi + v\sin\psi\right) + u\left(\left(u^2 + v^2\right) C_2 - \frac{\omega^2 - \omega_0^2}{2\omega}\right), \tag{11.25}$$

with the parameters

$$C_1 = \frac{I_3 (2n_0 - 1)\chi^{(2)}\omega_0^2}{3 I_2 n_0^2 \omega^3}\,, \tag{11.26}$$

$$C_2 = \frac{3\omega_0^2 I_4 \chi^{(3)} (2n_0 - 3)}{8\omega I_2 n_0^2} + \frac{\omega_0^2 I_3^2 \left(\chi^{(2)}\right)^2}{6\omega^3 I_2^2 n_0^4} \left[3 n_0 \omega^2 (3 - n_0) - 2\omega_0^2 (5 + n_0(2n_0 - 5))\right]. \tag{11.27}$$

Equations (11.24) and (11.25) are easily recognized as describing the same type of system as eqns (3.2) and (3.3) for $\eta = 0$ and $F_0 = 0$. We have therefore approximated the dynamics of a single optical resonator mode with the equations for a parametric oscillator.

For an optical ring resonator made from lithium niobate ($LiNbO_3$), we select the material values $\chi^{(2)} = 2 \times 10^{-11}\,\mathrm{m\,V^{-1}}$, $\chi^{(3)} = 0\,\mathrm{m^2\,V^{-2}}$, $n_0 = 2.2$, and $Q = 10^6$ for a vacuum wavelength of $\lambda_{\mathrm{vac}} = 1.55\,\mu\mathrm{m} = n_0 \lambda_{\mathrm{opt}}$. We model the ring resonator as a Fabry–Pérot cavity (optical medium between two mirrors) with periodic boundary conditions, and thereby neglect the small curvature of the ring. Assuming a square waveguide with width $L_x = 1000\,\mathrm{nm}$ and height $L_y = 400\,\mathrm{nm}$, and a ring circumference of $C_{\mathrm{ring}} = 1000\,\lambda_{\mathrm{opt}} \approx 700\,\mu\mathrm{m}$ [159], we very roughly model the mode shape as

$$\phi_1 = e^{-4r_x^2/L_x^2} \times e^{-4r_y^2/L_y^2} \times \cos^2\left(2\pi z/\lambda_{\mathrm{opt}}\right) , \tag{11.28}$$

with z the direction along the cavity axis. Equation (11.28) can be evaluated numerically or analytically to obtain the geometrical factors I_1 to I_4.

11.4 Rescaling of the Numerical Values

The exercises suggested in this book contain numerical simulations that can be performed with the provided code examples. Numerical simulations are very similar to experiments in the sense that they can produce artifacts or errors in certain situations. To help the reader avoiding such problems as much as possible, we summarize here a few useful guidelines for working with the numerical code examples.

Inserting realistic parameter values directly into a simulation algorithm often results in numerical problems and unphysical predictions such as unbounded growth of the amplitude. As a general rule, always first run an algorithm in situations where you know what result to expect, and only then introduce unknown elements bit by bit. Use common sense to recognize unphysical situations. Check that the result does not depend on the density of calculated points in time. For instance, the result should be invariant with respect to doubling the number of steps but keeping the total time constant.

To avoid numerical issues stemming from very large or small parameter values, we often rescale the physical values of our system to obtain dimensionless parameters in an accessible numerical range. There are a number of methods that can be used to rescale equations [160]. Here, we briefly discuss two of them.

Let us begin with the example of a harmonic oscillator without an external force, cf. eqn (1.5). Here, all terms in the differential equation are linear in x, such that the observed phenomena are invariant with respect to the amplitude. We can therefore rescale the initial condition x_{ini} with an arbitrary number, $z_{\mathrm{ini}} = x_{\mathrm{ini}}A$, and will obtain a result with a correspondingly scaled displacement $z = xA$. Similarly, we define a dimensionless, scaled time $\tau = \omega_0 t$, which transforms our system effectively into one with an angular resonance frequency of 1.[3] The original amplitude and timescales can then be recovered after a calculation by multiplying the time and amplitude axes by $1/\omega_0$ and $1/A$, respectively. When an external force is included, the driving frequency in the scaled time frame is rescaled with the resonance frequency as $\Omega = \omega/\omega_0$, while the force amplitude becomes $\bar{F}_0 = F_0 A/\omega_0^2$ in order to yield the correct long-time limit oscillation amplitude on resonance, $z = \bar{F}_0 Q/m$.

The linear rescaling method in amplitude does not provide correct results when nonlinear terms are present. Following the procedure used in Ref. [94], we instead define the displacement in space according to $z = x\sqrt{\beta/\omega_0^2}$, while the time scaling

[3] Be aware that Q does not need to be rescaled, but the damping coefficient becomes $\bar{\Gamma} = \Gamma/\omega_0 = 1/Q$.

remains as $\tau = \omega_0 t$. This rescaling of time and space allows us to rewrite eqn (11.7) in the form

$$\ddot{z} + [1 - \lambda \cos(2\Omega t + \psi)]\, z + z^3 + \bar{\Gamma}\dot{z} + \bar{\eta} z^2 \dot{z} = \frac{\bar{F}_0}{m}\cos(\Omega t) \tag{11.29}$$

where $\Omega = \omega/\omega_0$, $\bar{\Gamma} = 1/Q$, $\bar{\eta} = \eta\omega_0/\beta$, and $\bar{F}_0 = F_0\sqrt{\beta/m^2/\omega_0^3}$. If force noise is applied to the system, the force noise PSD is rescaled as $\varsigma_d^2 = \bar{S}_f = S_f \beta m^{-2}\omega_0^{-5}$ and the fluctuating force as $\bar{\xi} = \xi\sqrt{\beta/m^2/\omega_0^3}$. To obtain the true temporal and spatial evolution of the system, the results of the simulation are scaled back into their original dimensions by interpreting the time axis as $t = \tau/\omega_0$ and the displacement as $x = z\sqrt{\omega_0^2/\beta}$. The same rescaling procedure can be employed to rescale Hamiltonians, especially for the purpose of simulating quantum systems.

Chapter summary

- In Chapter 11, we learn how to apply all of the theoretical knowledge to realistic physical systems.
- The first example is a nanomechanical resonator made from a single carbon nanotube.
- The second example is a nonlinear electrical resonator built from a resistor, a coil inductor, and a nonlinear capacitance diode.
- The third example is a nonlinear optical circuit defined in a ring-shaped waveguide.
- We also include a brief explanation on how values much larger or smaller than 1 can be rescaled to improve the fidelity of numerical simulations.

Exercises

Just one task:

11.1 Choose a concrete physical system with realistic numerical values. Demonstrate the effects observed in this book on your own system in graphs with real units.

References

[1] M. Faraday, *Philosophical Transactions of the Royal Society of London* **121**, 299 (1831).

[2] F. Melde, *Annalen der Physik* **185**, 193 (1860).

[3] J. R. Sanmartín, *American Journal of Physics* **52**, 937 (1984).

[4] L. Rayleigh, *The London, Edinburgh, and Dublin Philosophical Magazine and Journal of Science* **15**, 229 (1883).

[5] E. F. W. Alexanderson, US Patent 1328797A (1915).

[6] G. Caryotakis, *Physics of Plasmas* **5**, 1590 (1998).

[7] A. Ashkin, J. C. Cook, W. H. Louisell, and C. F. Quate, US Patent 2958001A (1959).

[8] H. T. Closson, US Patent 3169226A (1962).

[9] H. Heffner and G. Wade, *Journal of Applied Physics* **29**, 1321 (1958).

[10] P. Penfield and R. P. Rafuse, *Varactor Applications*, MIT Press, 1962.

[11] L. Kuzmin, K. Likharev, V. Migulin, and A. Zorin, *IEEE Transactions on Magnetics* **19**, 618 (1983).

[12] J. R. Tucker and M. J. Feldman, *Reviews of Modern Physics* **57**, 1055 (1985).

[13] B. Yurke et al., *Physical Review* A **39**, 2519 (1989).

[14] M. A. Castellanos-Beltran and K. W. Lehnert, *Applied Physics Letters* **91**, 083509 (2007).

[15] T. Yamamoto et al., *Applied Physics Letters* **93**, 042510 (2008).

[16] A. Kamal, A. Marblestone, and M. Devoret, *Physical Review* B **79**, 184301 (2009).

[17] B. Abdo, A. Kamal, and M. Devoret, *Physical Review* B **87**, 014508 (2013).

[18] C. Eichler and A. Wallraff, *EPJ Quantum Technology* **1**, 2 (2014).

[19] M. Simoen et al., *Journal of Applied Physics* **118**, 154501 (2015).

[20] A. Roy and M. Devoret, *Comptes Rendus Physique* **17**, 740 (2016), Quantum microwaves / Micro-ondes quantiques.

[21] J. D. Teufel et al., *Nature* **475**, 359 (2011).

[22] D. Rugar and P. Grütter, *Physical Review Letters* **67**, 699 (1991).

[23] W. Zhang, R. Baskaran, and K. L. Turner, *Sensors and Actuators A: Physical* **102**, 139 (2002).

[24] R. B. Karabalin, S. C. Masmanidis, and M. L. Roukes, *Applied Physics Letters* **97**, 183101 (2010).

[25] R. B. Karabalin et al., *Physical Review Letters* **106**, 094102 (2011).

[26] L. G. Villanueva et al., *Nano Letters* **11**, 5054 (2011), PMID: 22007833.

[27] A. Eichler, J. Chaste, J. Moser, and A. Bachtold, *Nano Letters* **11**, 2699 (2011), PMID: 21615135.
[28] I. Mahboob, K. Nishiguchi, H. Okamoto, and H. Yamaguchi, *Nature Physics* **8**, 387 (2012).
[29] T. Faust, J. Rieger, M. J. Seitner, J. P. Kotthaus, and E. M. Weig, *Nature Physics* **9**, 485 (2013).
[30] H. Okamoto et al., *Nature Physics* **9**, 480 (2013).
[31] M. Frimmer and L. Novotny, *American Journal of Physics* **82**, 947 (2014).
[32] J. Gieseler, M. Spasenović, L. Novotny, and R. Quidant, *Physical Review Letters* **112**, 103603 (2014).
[33] M. Aspelmeyer, T. J. Kippenberg, and F. Marquardt, *Reviews of Modern Physics* **86**, 1391 (2014).
[34] J. Gieseler, B. Deutsch, R. Quidant, and L. Novotny, *Physical Review Letters* **109**, 103603 (2012).
[35] M. Poot, K. Y. Fong, and H. X. Tang, *Physical Review* A **90**, 063809 (2014).
[36] I. Mahboob, H. Okamoto, K. Onomitsu, and H. Yamaguchi, *Physical Review Letters* **113**, 167203 (2014).
[37] R. E. Slusher, L. W. Hollberg, B. Yurke, J. C. Mertz, and J. F. Valley, *Physical Review Letters* **55**, 2409 (1985).
[38] G. Breitenbach, S. Schiller, and J. Mlynek, *Nature* **387**, 471 (1997).
[39] E. E. Wollman et al., *Science* **349**, 952 (2015).
[40] J.-M. Pirkkalainen, E. Damskägg, M. Brandt, F. Massel, and M. A. Sillanpää, *Physical Review Letters* **115**, 243601 (2015).
[41] F. Lecocq, J. B. Clark, R. W. Simmonds, J. Aumentado, and J. D. Teufel, *Physical Review* X **5**, 041037 (2015).
[42] C. M. Caves, *Physical Review* D **23**, 1693 (1981).
[43] T. Eberle et al., *Physical Review Letters* **104**, 251102 (2010).
[44] H. Yu et al., *Nature* **583**, 43 (2020).
[45] M. C. Lifshitz, R. Cross, *Nonlinear Dynamics of Nanomechanical and Micromechanical Resonators*, pages 1–52, Wiley-VCH, 2009.
[46] J. R. Paulling, Parametric rolling of ships–then and now, in *Contemporary Ideas on Ship Stability and Capsizing in Waves*, pages 347–360, Springer, 2011.
[47] J. H. Traschen and R. H. Brandenberger, *Physical Review* D **42**, 2491 (1990).
[48] D. G. Figueroa and F. Torrenti, *Journal of Cosmology and Astroparticle Physics* **2017**, 001 (2017).
[49] M. I. Dykman, C. M. Maloney, V. N. Smelyanskiy, and M. Silverstein, *Physical Review* E **57**, 5202 (1998).
[50] E. Goto, *Proceedings of the IRE* **47**, 1304 (1959).
[51] J. v. Neumann, US Patent 2815488 (1959).
[52] M. Dykman, *Fluctuating Nonlinear Oscillators*, Oxford University Press, 2012.
[53] M. Marthaler and M. I. Dykman, *Physical Review Letters* A **73**, 042108 (2006).
[54] D. Ryvkine and M. I. Dykman, *Physical Review* E **74**, 061118 (2006).
[55] I. Mahboob and H. Yamaguchi, *Nature Nanotechnology* **3**, 275 (2008).

[56] J. F. Rhoads and S. W. Shaw, *Applied Physics Letters* **96**, 234101 (2010).
[57] I. Mahboob, E. Flurin, K. Nishiguchi, A. Fujiwara, and H. Yamaguchi, *Nature Communications* **2**, 198 (2011).
[58] Z. Wang, A. Marandi, K. Wen, R. L. Byer, and Y. Yamamoto, *Physical Review* A **88**, 063853 (2013).
[59] A. Marandi, Z. Wang, K. Takata, R. L. Byer, and Y. Yamamoto, *Nature Photonics* **8**, 937 (2014).
[60] Z. Lin et al., *Nature Communications* **5**, 4480 (2014).
[61] D. A. Czaplewski et al., *Physical Review Letters* **121**, 244302 (2018).
[62] M. I. Dykman, C. Bruder, N. Lörch, and Y. Zhang, *Physical Review* B **98**, 195444 (2018).
[63] A. Grimm et al., *Nature* **584**, 205 (2019).
[64] Z. Wang et al., *Physical Review* X **9**, 021049 (2019).
[65] J. M. Miller, D. D. Shin, H.-K. Kwon, S. W. Shaw, and T. W. Kenny, *Physical Review Applied* **12**, 044053 (2019).
[66] J. J. Hopfield, *Proceedings of the National Academy of Sciences* **79**, 2554 (1982).
[67] R. Rojas, *Neural Networks*, Springer, 1996.
[68] E. Ising, *Zeitschrift für Physik* **31**, 253 (1925).
[69] T. Inagaki et al., *Science* **354**, 603 (2016).
[70] H. Goto, K. Tatsumura, and A. R. Dixon, *Science Advances* **5**, eaav2372 (2019).
[71] S. E. Nigg, N. Lörch, and R. P. Tiwari, *Science Advances* **3**, e1602273 (2017).
[72] N. Mohseni, P. L. McMahon, and T. Byrnes, *Nature Reviews Physics*, 1 (2022).
[73] I. Mahboob, C. Froitier, and H. Yamaguchi, *Applied Physics Letters* **96**, 213103 (2010).
[74] A. Leuch et al., *Physical Review Letters* **117**, 214101 (2016).
[75] Ž. Nosan, P. Märki, N. Hauff, C. Knaut, and A. Eichler, *Physical Review* E **99**, 062205 (2019).
[76] T. L. Heugel, M. Oscity, A. Eichler, O. Zilberberg, and R. Chitra, *Physical Review* **123**, 124301 (2019).
[77] M. Frimmer et al., *Physical Review Letters* **123**, 254102 (2019).
[78] W. Lechner, P. Hauke, and P. Zoller, *Science Advances* **1** (2015).
[79] I. Mahboob, H. Okamoto, and H. Yamaguchi, *Science Advances* **2**, e1600236 (2016).
[80] T. Inagaki et al., *Nature Photonics* **10**, 415 (2016).
[81] H. Goto, Scientific Reports **6**, 21686 (2016).
[82] G. Csaba, T. Ytterdal, and W. Porod, Neural network based on parametrically-pumped oscillators, in *2016 IEEE International Conference on Electronics, Circuits and Systems (ICECS)*, pages 45–48, 2016.
[83] S. Puri, S. Boutin, and A. Blais, *npj Quantum Information* **3**, 18 (2017).
[84] H. Goto, Z. Lin, and Y. Nakamura, *Scientific Reports* **8**, 7154 (2018).
[85] R. Rota, F. Minganti, C. Ciuti, and V. Savona, *Physical Review Letters* **122**, 110405 (2019).

[86] T. L. Heugel, O. Zilberberg, C. Marty, R. Chitra, and A. Eichler, *Physical Review Research* **4**, 013149 (2022).
[87] L. Landau and E. Lifshitz, *Mechanics*, Butterworth-Heinemann, 1976.
[88] J. A. Richards, *Analysis of Periodically Time-Varying Systems*, Springer-Verlag Berlin, Heidelberg, 1983.
[89] A. J. Leggett et al., *Reviews of Modern Physics* **59**, 1 (1987).
[90] F. M. Fernández, *European Journal of Physics* **39**, 045005 (2018).
[91] R. A. Rand, *Lecture Notes on Nonlinear Vibrations*, Cornell University, 2005.
[92] D. Jordan and P. Smith, *Nonlinear Ordinary Differential Equations: An Introduction for Scientists and Engineers*, Oxford University Press, 2007.
[93] C. Holmes and P. Holmes, *Journal of Sound and Vibration* **78**, 161 (1981).
[94] L. Papariello, O. Zilberberg, A. Eichler, and R. Chitra, *Physical Review* E **94**, 022201 (2016).
[95] M. Krack and J. Gross, *Harmonic Balance for Nonlinear Vibration Problems*, volume 1, Springer, 2019.
[96] J. Košata, J. del Pino, T. L. Heugel, and O. Zilberberg, *SciPost Physics Codebases*, 006 (2022).
[97] J. Woo and R. Landauer, *IEEE Journal of Quantum Electronics* **7**, 435 (1971).
[98] H. B. Callen and T. A. Welton, *Physical Review* **83**, 34 (1951).
[99] R. Z. Khasminskii, *Theory of Probability & Its Applications* **11**, 390 (1966).
[100] J. Roberts and P. Spanos, *International Journal of Non-Linear Mechanics* **21**, 111 (1986).
[101] K. Schulten and I. Kosztin, *Theory and Application of Mathieu Functions*, lecture notes, University of Illinois at Urbana-Champaign, 2000.
[102] L. C. Evans, *An Introduction to Stochastic Differential Equations*, American Mathematical Society, 2013.
[103] H. Risken, Fokker-Planck equation, in *The Fokker-Planck Equation*, pages 63–95, Springer, 1996.
[104] T. D. Frank, *Nonlinear Fokker-Planck Equations: Fundamentals and Applications*, Springer Science & Business Media, 2005.
[105] A. Vinante and P. Falferi, *Physical Review Letters* **111**, 207203 (2013).
[106] J. S. Huber et al., Phys. Rev. X **10**, 021066 (2020).
[107] M. Soriente, T. L. Heugel, K. Omiya, R. Chitra, and O. Zilberberg, *Physical Review Research* **3**, 023100 (2021).
[108] P. Hänggi, P. Talkner, and M. Borkovec, *Reviews of Modern Physics* **62**, 251 (1990).
[109] H. Chan, M. Dykman, and C. Stambaugh, *Physical Review Letters* **100**, 130602 (2008).
[110] D. Hälg et al., *Physical Review Letters* **128**, 094301 (2022).
[111] M. Frimmer, J. Gieseler, T. Ihn, and L. Novotny, *Journal of the Optical Society of America* B **34**, C52 (2017).
[112] P. Forn-Díaz, L. Lamata, E. Rico, J. Kono, and E. Solano, *Reviews of Modern Physics.* **91**, 025005 (2019).

[113] A. Olkhovets, D. W. Carr, J. M. Parpia, and H. G. Craighead, Non-Degenerate Nanomechanical Parametric Amplifier, in *Technical Digest. MEMS 2001. 14th IEEE International Conference on Micro Electro Mechanical Systems (Cat. No. 01CH37090)*, pages 298–300, 2001.

[114] E. Kenig et al., *Physical Review Letters* **108**, 264102 (2012).

[115] F. Sun, X. Dong, J. Zou, M. I. Dykman, and H. B. Chan, *Nature Communications* **7**, 1 (2016).

[116] L. Bello, M. Calvanese Strinati, E. G. Dalla Torre, and A. Pe'er, *Physical Review Letters* **123**, 083901 (2019).

[117] W. P. Robins, *Phase Noise in Signal Sources*, Institution of Engineering and Technology, London, 1984.

[118] F. Marquardt and S. M. Girvin, *Physics* **2**, 40 (2009).

[119] E. Buks and M. L. Roukes, *Journal of Microelectromechanical Systems* **11**, 802 (2002).

[120] R. Lifshitz and M. C. Cross, *Physical Review* B **67**, 134302 (2003).

[121] M. Calvanese Strinati, L. Bello, E. G. Dalla Torre, and A. Pe'er, *Physical Review Letters* **126**, 143901 (2021).

[122] C. Cohen-Tannoudji, B. Diu, and F. Laloe, *Quantum Mechanics Vol. 1*, Wiley, 1991.

[123] L. Mandelstam and I. Tamm, The Uncertainty Relation Between Energy and Time in Non-Relativistic Quantum Mechanics, in *Selected Papers*, pages 115–123, Springer, 1991.

[124] W. Demtröder, *Laser Spectroscopy, Vol. 1*, Springer, 2008.

[125] W. P. Robins, *Phase Noise in Signal Sources: Theory and Applications*, volume 9, IET, 1984.

[126] D. Gabor, *Journal of the Institution of Electrical Engineers, Part III: Radio and Communication Engineering* **93**, 429 (1946).

[127] M. Hall, *First Break* **24** (2006).

[128] J. Von Neumann, *Mathematical Foundations of Quantum Mechanics*, Princeton University Press, 2018.

[129] H. M. Wiseman and G. J. Milburn, *Quantum Measurement and Control*, Cambridge University Press, 2009.

[130] A. D. O'Connell et al., *Nature* **464**, 697 (2010).

[131] L. De Broglie, *Nature* **112**, 540 (1923).

[132] J. R. Johansson, P. D. Nation, and F. Nori, *Computer Physics Communications* **183**, 1760 (2012).

[133] E. Wigner, *Physical Review* **40**, 749 (1932).

[134] S. Haroche and J.-M. Raimond, *Exploring the Quantum: Atoms, Cavities, and Photons*, Oxford University Press, 2006.

[135] A. Royer, *Physical Review* A **15**, 449 (1977).

[136] H.-P. Breuer and F. Petruccione, *The Theory of Open Quantum Systems*, Oxford University Press, 2003.

[137] C. Gardiner and P. Zoller, *Quantum Noise: A Handbook of Markovian and Non-Markovian Quantum Stochastic Methods with Applications to Quantum Optics*, Springer Science & Business Media, 2004.

[138] D. A. Steck, *Quantum and Atom Optics*, available online at http://steck.us/teaching, 2007.

[139] D. Manzano, *AIP Advances* **10**, 025106 (2020).

[140] W. P. Bowen and G. J. Milburn, *Quantum Optomechanics*, CRC Press, 2015.

[141] T. Pearsall, *Quantum Photonics*, 2nd edition, Springer, 2020.

[142] U. Leonhardt, *Measuring the Quantum State of Light*, vol. 22, Cambridge University Press, 1997.

[143] J. Košata, A. Leuch, T. Kästli, and O. Zilberberg, *Physical Review Research* **4**, 033177 (2022).

[144] N. Vitanov et al., *Optics Communications* **199**, 117 (2001).

[145] A. Eichler et al., *Nature Nanotechnology* **6**, 339 (2011).

[146] T. L. Heugel, M. Biondi, O. Zilberberg, and R. Chitra, *Physical Review Letters* **123**, 173601 (2019).

[147] M. Born and V. Fock, *Zeitschrift für Physik* **51**, 165 (1928).

[148] V. Sazonova et al., *Nature* **431**, 284 (2004).

[149] E. A. Laird et al., *Reviews of Modern Physics* **87**, 703 (2015).

[150] A. Eichler, M. del Álamo Ruiz, J. A. Plaza, and A. Bachtold, *Physical Review Letters* **109**, 025503 (2012).

[151] A. Eichler, J. Moser, M. I. Dykman, and A. Bachtold, *Nature Communications* **4**, 2843 (2013).

[152] J. Moser, A. Eichler, J. Güttinger, M. I. Dykman, and A. Bachtold, *Nature Nanotechnology* **9**, 1007 (2014).

[153] C. Müller, J. H. Cole, and J. Lisenfeld, *Reports on Progress in Physics* **82**, 124501 (2019), IOP Publishing.

[154] M. H. Devoret and R. J. Schoelkopf, *Science* **339**, 1169 (2013).

[155] C. Eichler, Y. Salathe, J. Mlynek, S. Schmidt, and A. Wallraff, *Physical Review Letters* **113**, 110502 (2014).

[156] S. Puri et al., *Physical Review* X **9**, 041009 (2019).

[157] R. W. Boyd, *Nonlinear Optics*, Academic Press, 2020.

[158] K. Okamoto, *Fundamentals of Optical Waveguides*, Elsevier, 2021.

[159] M. Zhang, C. Wang, R. Cheng, A. Shams-Ansari, and M. Lončar, *Optica* **4**, 1536 (2017).

[160] H. P. Langtangen and G. K. Pedersen, *Scaling of Differential Equations*, Springer Nature, 2016.

Subject Index

adiabatric theorem, 152
amplitudes
  corrections, Duffing resonator, 23
  linear rescaling, 166–7
  response of the driven resonator, 8, 8*f*
antisymmetric contribution, resonator potential, 27, 27*f*
Arnold tongues, 16–17, 16*f*, 35
  boundary determination, 35–6, 35*f*
  boundary of, 40
  definition, 98
  lowest, 32
  strong coupling & weak pumping, 105
  weak coupled parametric oscillations, 101*f*
  width of, 100
asymmetric phase configurations, 104
attractors, hopping between, 71–2
autocovariance, 61
  force noise response, 48*f*, 49–50, 49*f*
  Fourier transformation, 48
averaging method, quartic potential, 23–5, 29
averaging theorem, 25

bifurcation points, 27
Bloch equations, 88
Boltzmann constant, 46

carbon nanotube, 159*f*
cavity optomechanics, 3
characteristic exponents, 76
  parametric pumping above threshold, 69–71
characteristic multipliers, Floquet theory, 14
classical Fourier analysis, 114–15, 115*f*
classical parametric oscillator, 151
classical squeezing, 65, 65*f*, 76
coherent state, 121
coherent states, quantum superposition, 156
correspondence principle, 139
coupled harmonic resonators, 79–97
  alternative types of coupling, 91–5
  nondegenerate three-wave mixing, 86–91
  static coupling, 79–86
coupled parametric oscillations, 98–112
  $N = 2$ examples, 100–8, 101*f*, 106*f*
  N, equations for, 98–100
  networks with N 2, 108–10
coupled quantum parametric oscillators, 153–4
coupling
  dissipative *see* dissipative coupling
  parametric modulation, 90*f*, 96
  quadratic force, 79n
  strong *see* strong coupling
  ultrastrong, 88*f*, 89
cross-Kerr terms, 154
cubic potential
  Duffing resonator, 27–8
  second-order averaging, 28

damping below threshold, nonlinear parametric resonator, 33–5
degenerate parametric pumping, 32–45
  nonlinear parametric resonator, 32–41
degenerate three-wave mixing, 42
density matrix, 130
  Fock states, 123
detuning, static coupled harmonic resonators, 85–6
differential equation, matrix formulation, 10–11
Dirac's notation, quantum harmonic oscillator, 119–20
disappearance of hysteresis, 76
discrete double-sided PSD, 52
discrete signals, force noise, 51–2
discrete transition matrix, Floquet theory, 14
dissipative coupling, 92–3, 93*f*
  alternative coupled harmonic resonators, 92–3
driven Duffing resonator, 25–6
  nonlinear damping, 33
driven quantum resonator, quantum systems, open-to-closed, 136–40
driven resonator response, 7–8
Duffing nonlinearity, 20, 29, 143n, 162
  cubic potential *vs.*, 27
Duffing oscillator, stationary response of, 26–7
Duffing resonator, 20–31
  cubic potential, 27–8
  driven *see* driven Duffing resonator
  parametric pumping, 42
  quartic potential, 20–7, 20*f*

effective rotating frame Hamiltonian, 44

electrical resonators, 161–3, 161*f*
equations of motion (EOM)
  Hamilton's equation of motion, 18
  Liouville's equation of motion, 132–4
  Newton's equation of motion, 5–7, 18
  semi-classical equation of motion, 144–5
equipartition theorem, 46, 61
exceptional point, 7n
expectation value, 140
experimental systems, 159–68
  electrical resonators, 161–3, 161*f*
  mechanical resonators, 159*f*, 160–1
  numerical value rescaling, 166–7
  optical resonators, 163–6
external driving, quantum resonator, 138–40, 139*f*, 140*f*
external force, parametric pumping, 152–3, 154*f*

Fabry-Pérot cavity, 165–6
fast Fourier transform, 69*f*
finite time resolution, 47n
first-order averaging, quartic potential, 25–7, 26*f*, 27*f*
Floquet theory, 13–17, 18
fluctuation–dissipation theorem (FDT), 52–7
  nonrotating frame, 53–4
  rotating frame, 54–6
Fock ladder, 145–6, 149
Fock level broadening, 145–6
Fock spectrum, 130
Fock states, 118
  density matrix, 123
  maximum amplitude, 126–7
  notation, 120
  Poisson distribution, 121
Fokker–Planck equation, 75, 76
  noise-driven harmonic resonator, 58*f*, 59
  parametric resonators with force noise, 64
  probability distribution approach, 58–60
  steady state, 58*f*
Fokker–Planck formalism, 59
force noise, 46–51, 47*f*
  discrete signals, 51–2
  parametric resonators with *see* parametric resonators with force noise
  response as function of time, 48–50, 48*f*
  rotating frame, 56–7, 56*f*
  spectral response, 50–1
  thermal *see* thermal force noise
Fourier analysis, classical, 114–15, 115*f*
Fourier transformation
  autocovariance, 48
  fast, 69*f*
  response of the driven resonator, 7
  Wiener–Khinchin theorem, 50
four-wave mixing, quartic potential, 21
frequency difference driving, nondegenerate three-wave mixing, 86–90, 87*f*, 88*f*
frequency sum driving, nondegenerate three-wave mixing, 90–1, 90*f*
full-width-at-half maximum (FWHM), 8

general Hamiltonian, quantum parametric oscillator, 143–5
generating functions, 24
ground state energy, quantum harmonic oscillator, 115–16
gyromagnetic ratio, 148

Hamiltonian
  effective rotating frame, 44
  general, 143–5
  Newton's equation of motion, 5
Hamiltonian quantum evolution, 125
Hamilton's equation of motion, 5, 18, 57
harmonic balance, 24n
harmonic oscillators, 166
  example of, 6*f*
  parametric pumping, 34*f*
  solution of, 22
harmonic resonators, 5–19
  coupled *see* coupled harmonic resonators
harmonics, Duffing resonator, 23
Heisenberg uncertainty principle, 130
  quantum harmonic oscillator, 114–15
Hermite polynomials, 117–18
Hilbert space, 119
Hill equation, undampend, 12
homogeneous dissipative, 6–7
hopping between attractors, 76
  parametric pumping above threshold, 71–2
Husimi-Q distribution, 123, 127, 130
hysteresis, 27
  disappearance of, 76

in-phase component, 95
integral equation, matrix formulation, 10–11
Ising Hamiltonians, 108–9, 111
Îto process, 47

Jacobian matrix, 69–71
Josephson junctions, electrical resonators, 162
Josephson parametric amplifiers, 3

Kerr term, 143n
Klystron amplifiers, 2–3
Krylov–Bogolyubov averaging method, 23
  fluctuation–dissipation theorem, 54–5

ladder operators, 120
Lindblad master equation, 132–4, 141, 154–5, 156
linear rescaling, amplitude, 166–7
Liouville's equation of motion, 132–4
Liouville's theorem, probability distribution approach, 57–8

Mag Amp, 2–3
Mandelshtam–Tamm inequality, 114

master equation, quantum harmonic oscillator, 122
Mathieu equation, Meissner equation *vs.*, 14–15, 16*f*
matrix formulation, 8–11
  equation solving, 9
MAX-CUT problem, 4
mechanical resonators, 159*f*, 160–1
Meissner equation, 18
  Mathieu equation *vs.*, 14–15, 16*f*
mixed states, quantum harmonic oscillator, 122
modified bifurcation topology, 41
multistability
  parametric resonators with force noise, 63–4, 63*f*
  probability density picture in, 72–4, 72*f*, 73*f*, 74*f*

natural frequency, 6
near-identity transformation, 24–5
Newton's equation of motion, 5–7, 18
noise-driven harmonic resonator, Fokker–Planck equation, 58*f*, 59
nondegenerate three-wave mixing, coupled harmonic resonators, 86–91
nonlinear coupling, alternative coupled harmonic resonators, 93–5, 94*f*
nonlinear parametric resonator, degenerate parametric pumping, 32–41
nonlinear quantum parametric oscillator, 145–8, 146*f*
nonrotating frame, fluctuation–dissipation theorem, 53–4
normal mode coupling, static coupled harmonic resonators, 84*f*, 85–6
normal mode picture, static coupled harmonic resonators, 81–5
number operators, 120
numerical value rescaling, 166–7

optical power spectrum, 94*f*
optical resonators, 163–6
optical ring resonators, 163*f*, 165–6
optomechanics, 94*f*
out-of-phase component, 95

parametric amplification, 34–5
  below threshold, parametric resonators with force noise, 64–8, 64*f*
parametric instability threshold, 33–4
parametric modulation, 11–13, 13*f*
  coupling, 90*f*, 96
parametric oscillators
  classical, 151
  coupled *see* coupled parametric oscillations
  coupled quantum, 153–4
  historical example, 2
parametric phase logic, 3–4
parametric phase symmetry, 44
parametric pumping, 155*f*
  degenerate *see* degenerate parametric pumping
  external force, 152–3, 154*f*
  harmonic oscillator, 34*f*
  historical example, 2
  quantum systems and, 148–52, 150*f*
  strong *see* strong parametric pumping
  sweeping, 103
  three-wave mixing, 42–3
  weak *see* weak parametric pumping
parametric pumping above threshold, 35–6, 68–74, 68*f*, 69*f*
  characteristic exponent method, 68*f*, 69–71, 70*f*
  hopping between attractors, 71–2, 71*f*
  multistability in probability density, 72–4, 72*f*, 73*f*, 74*f*
parametric resonators with force noise, 63–78
  hierarchy of relevant timescales, 74–5
  multistability solutions, 63–4, 63*f*
  parametric amplification below threshold, 64–8, 65*f*
  quasi-stable solutions, 63–4
parametric symmetry breaking, 40–1, 41*f*
parametric, term definition, 2
parametric threshold, 44
parametron, 3
Parseval's theorem, 51, 52, 61
  fluctuation–dissipation theorem, 54
particle-wave duality, 117
periodicity doubling, nonlinear parametric resonator, 32–3
phase response, response of the driven resonator, 8, 8*f*
phase space, representations of, 127
phenomena, quantum parametric oscillator, 145–53
Planck constant, reduced, 114
Planck-Einstein relation, 118
Poincaré–Linstedt method, 26, 26*f*, 29
  quartic potential, 20*f*, 22–3
Poisson distribution, Fock states, 121
power broadening, 147
power definition, 51n
power spectral density (PSD), 61
probability amplitude, 117
probability density picture, multistability in, 72–4
probability distribution approach, 57–60
  Fokker–Planck equation, 58–60
  Liouville's theorem, 57–8
PSD (power spectral density), 61

quadratic force, coupling, 79n
quantum harmonic oscillator, 113–31
  from classical, 113–16, 113*f*
  first to second quantization, 116–22

quantum state representation *see* quantum state representation
quantum mechanics, wave mechanics, 119–20
quantum parametric oscillators, 143–58
general Hamiltonian, 143–5
nonlinear, 145–8, 146*f*
phenomena, 145–53
quantum parametric oscillators, coupled, 153–4
quantum resonators, driven, 136–40
quantum state representation, 123–9
definitions, 123–5
visualizations of, 124*f*, 125–7, 125*f*, 127*f*, 128*f*, 129, 129*f*
quantum superposition of coherent states, 156
quantum systems
open-to-closed, 132–42
driven quantum resonator, 136–40
quartic potential
Duffing resonator, 20–7, 20*f*
first-order averaging, 25–7
quasi-potential barriers, 63, 63*f*
quasi-potential picture, 44
stability and, 36–40
quasi-stable solutions, parametric resonators with force noise, 63–4
QuTiP, 123

Rabi oscillations, 146–8, 147*f*, 156
reduced Planck constant, 114
reduced system, 133
resonant parametric modulation, 12–13
resonator displacement, 162
resonator potential, antisymmetric contribution, 27, 27*f*
rotating frame, 26
fluctuation–dissipation theorem, 54–6
force noise, 56–7, 56*f*
rotating phase space, driven quantum resonator, 136–8
rotating-wave approximation (RWA), 137
RWA (rotating-wave approximation), 137

Schrödinger equation, 117–18, 130
matrix form, 146
mixed states, 122
quantum harmonic oscillator, 116–18, 117*f*, 119*f*
second quantization, quantum harmonic oscillator, 120–1
second-order averaging, cubic potential, 28
secular perturbation technique, 23
self-Kerr terms, 154
semi-classical equation of motion, 144–5
sideband control, 95
slow-flow equations
N, equations for coupled parametric oscillations, 99–100
parametric resonators with force noise, 64
spectral response, force noise, 50–1
spontaneous time-translation symmetry, 68
squeezing, 64–5
classical, 65, 65*f*, 76
stability analysis
matrix formulation, 9–10
quasi-potential picture, 38*f*, 39
state transition matrix, 18
static coupling, harmonic resonators, 79–86
stationary response, Duffing oscillator of, 26–7
strong coupling
limit, 85
strong parametric pumping and, coupled parametric oscillations, 106–8, 106*f*, 108*f*
weak parametric pumping and, coupled parametric oscillations, 105, 107*f*
strong parametric pumping, 111
strong coupling and, coupled parametric oscillations, 106–8, 106*f*, 108*f*
sweeping, parametric pumping, 103
symmetric phase configurations, 104
symmetry phase space, 105, 105*f*

thermal equilibrium, 46
thermal force noise, 46
3-dB limit, 66
3-dB limit, parametric resonators with force noise, 66–8
three-wave mixing, 27–8
degenerate, 42
parametric pumping, 42–3
time function, force noise response as, 48–50
timescale hierarchy, parametric resonators with force noise, 74–5
time-translation symmetry breaking bifurcation, 44
total fluctuating displacement, 67

ultrastrong coupling regime, 88*f*, 89
undamped harmonic resonator, Floquet theory, 14
undamped Hill equation, parametric modulation, 12
undamped system, Floquet theory, 15–16

van der Pol transformation, 24
variational ansatz, 10
von Neumann master equation, 122, 130, 133, 141

wave mechanics, quantum mechanics, 119–20
weak parametric coupling, 111
strong pumping and, coupled parametric oscillations, 101–5, 101*f*, 103*f*, 105*f*
weak pumping and, coupled parametric oscillations, 101, 101*f*
weak parametric pumping
coupled parametric oscillations, 101

strong coupling and, coupled parametric oscillations, 105, 107*f*
Wiener process, 47
Wiener–Khinchin theorem, 50, 51, 61
Wigner quasiprobability distribution, 123, 124–5, 128*f*, 129, 129*f*, 130
Wigner representations, 148
Wronskian matrices, 10, 18
Floquet theory, 15
fluctuation–dissipation theorem, 53
static coupled harmonic resonators, 81